Garfield County Libraries
Parachute Branch Library
244 Grand Valley Way
Parachute, CO 81635
(970) 285-9870 Fax (970) 285-7477
www.garfieldlibraries.org

Being with Animals

DOUBLEDAY

New York / London

Toronto / Sydney

Auckland

BARBARA J. KING

Being with
Animals

•   •   •   •   •   •   •

*Why We Are Obsessed with
the Furry, Scaly,
Feathered Creatures Who
Populate Our World*

DD

DOUBLEDAY

Copyright © 2010 by Barbara J. King

Published in the United States by Doubleday Religion, an imprint of the
Crown Publishing Group, a division of Random House, Inc., New York.
www.crownpublishing.com

DOUBLEDAY and the DD colophon are registered trademarks of
Random House, Inc.

"Little Dog's Rhapsody in the Night (Three)" on page 169 is from *The Truro
Bear and Other Adventures* by Mary Oliver, copyright © 2008 by Mary
Oliver. Reprinted by permission of Beacon Press.

Library of Congress Cataloging-in-Publication Data

King, Barbara J.
Being with animals : why we are obsessed with the furry, scaly, feathered
creatures who populate our world / Barbara J. King.—1st ed.
    p.   cm.
Includes bibliographical references and index.
1. Animal behavior—Popular works.   2. Emotions in animals—Popular works.
3. Human-animal relationships—Popular works.   I. Title.
QL751.K5175 2009
590—dc22
2009020030

ISBN 978-0-385-52363-9

Printed in the United States of America

Design by Ellen Cipriano

10 9 8 7 6 5 4 3 2 1

First Edition

To Bruce Springsteen and the E Street Band

*Who may never read these words, but who taught me that
all things are possible in concert*

"Let me see your hands . . ."

Contents

• • • •

Being with Animals

1

A Holy Procession of Animals

• • • •

Little Lamb, who made thee?
Dost thou know who made thee?

—William Blake, "The Lamb," 1789

*E*VERYONE COULD FEEL IT. Anticipation weighted the air as the
Saint Francis Day church service unfolded at the Cathedral of
St. John the Divine in New York City. Finally, the massive front door
swung open. Into the largest of all cathedrals in North America
flooded an early-autumn light, a light that illuminated the creature
that walked first through the door: a camel, head held high, its single
hump garlanded with flowers.

Anticipation turned to reverence, and twelve hundred people
turned their gaze as one. Behind the camel and its white-robed caretaker came a royal yak, a tundra reindeer, a baby wallaby and a baby
ape, and a black sheep named Marvin. A giant tortoise named Oscar
rode on two pillows arranged on a wagon. Birds of prey, and a host of
smaller creatures from bunnies to hermit crabs, were carried or pulled
on carts.

Slowly, with calm and dignity, the Procession of Animals made its
way to the front of the church. There the animals were blessed by

waiting clergy. As much as the individual animals themselves, it was the animal-human relationship that was celebrated. The choir sang, and young girls lined the aisles, waving colorful flags. As the animals slowly turned to move back up the aisle, the human congregants sang and moved in harmony with the music.

As I watched the ritual, I saw how the joys inherent in sharing the world with animals lighted people's faces and enriched their voices. Throughout the church, people bent forward to whisper a word to the dog by their side, or to re-settle into their carriers a cat or a rabbit. People not only came to the ceremony in the thousands, they also brought their own animals along. When the service concluded, people and animals walked, two by two, into the cathedral's garden so that clergy could bless each pet with loving words and touch.

The blessing ceremony at St. John's is the most elaborate and famous of any in the world, but many smaller ceremonies in other towns and cities take place on the first Sunday in October, a day set aside to remember the patron saint of animals. People attend not to watch passively but to participate actively, to bring into alignment and harmony their love of animals and their love of God.

Rituals like this, whether focused on the Christian God or some other modern God or gods, mark an emotional connection between animals and people that stretches far back into human prehistory. Our species, *Homo sapiens, became* human by being with animals.

Deep inside a cave in prehistoric France, early *Homo sapiens* people gathered in near darkness. Artists in the group had adorned the walls with pigments, rich reds and jet blacks, in order to create spectacular animal images. Now the group assembled in dim light, singing and moving rhythmically together.

As they lost themselves in the pulse and the beat, some people began to experience a slightly altered consciousness and a heightened connection with the living creatures whose representations graced the walls. Hunters felt at one with the animals they would stalk the next day. A skilled healer stared at a half-man, half-bird image cruder than the others. Feeling the first stirrings of a transformation, he knew he would soon be released from his earthly moorings and in contact with otherworldly creatures and forces.[1]

In prehistoric Turkey, at the village of Catalhöyük around 8,000 years ago, a man was buried together with a lamb. The bodies, one person and one animal, were kept slightly separate in death by an unusual, contorted position of the lamb and by the placement of a mat or blanket between the two. Yet, in a place where animals were routinely domesticated but not usually buried, the lamb was placed in a gravesite used traditionally for human ancestors: a pit dug beneath a house floor. In subsequent years, three other people were buried there as well.[2]

At around the same time in Israel, people at a place called Kfar HaHoresh constructed a large mosaic. The material used was not tile, but the carefully positioned bones of humans and gazelles. The image (when viewed from above) is the profile of an animal, perhaps a boar, an aurochs, or even a lion. Elsewhere at the site, a human skull was buried underneath the floor of a rectangular structure and just above a headless gazelle carcass.[3]

In ancient China, a hermit called Zhuangzi entered a game park and took aim at a magpie. Preoccupied with a cicada, the magpie did not notice Zhuangzi; neither the cicada nor a nearby preying mantis noticed the magpie. The magpie "swept down on its prey in high excitement and gobbled them both up." A feeling of compassion welled up in Zhuangzi: here in the certainty of death was the essence of life. For months, Zhuangzi felt depressed, but also enlivened by new thoughts: Life is about endless transformation, and death should not be feared; realizing this, Zhuangzi felt an "exhilarating freedom" that changed his life.[4]

In modern-day California, a small group of people shared an amazing encounter with a fifty-foot-long, fifty-ton whale. In the waters beyond San Francisco's Golden Gate Bridge, a female humpback whale became tangled in equipment used by crabbers, about twenty crab-pot ropes, each 240 feet long and with weights attached every sixty feet. Unable to free herself, and struggling to keep her blowhole above the waterline, the whale was in real peril.

When rescuers arrived by boat, they determined that the best hope of saving the creature would be to cut her free, working *in the water*. To swim and dive right near such a powerful—and frightened— sea mammal presented serious risk. The procedure to free the whale took an hour, but the mighty creature remained calm throughout, and emitted what rescuers described as a strange kind of vibration. Once freed, she did not bolt for the open sea. Rather, she circled her four rescuers in a way that struck them as joyful. She then approached each person and nudged each in turn. Diver James Moskito said later, "It felt to me like it was thanking us, knowing that it was free and that we had helped it. . . . When I was cutting the line going through the mouth, its eye was there winking at me, watching me. It was an epic moment of my life. . . . It was an amazing, unbelievable experience."[5]

These examples tell a fundamental truth about how we humans make sense of the world: we think and we feel through being with animals. Utterly unique to the human-animal realm is an emotional kind of mutual relating. Yes, the ever-shifting light on the walls of the ancient Grand Canyon or on the cypress trees swaying gently in a Tuscan breeze may move us profoundly. Yet neither the light, the canyon, nor the trees will ever actively engage with our admiring gaze, or with our emotions. They will never give back emotionally as animals can.

To feel that mutual kinship with another creature is a special experience, one that brings us into attunement with the whole world. It's a feeling deep in the chest, resonant in the heart: *I share with all creatures a way of being in this world. All animals, in their own ways, struggle to live, and* feel *their lives in different ways. I belong here in this world, with them.*

We know these things, and in a sense we may even take them for granted. After all, we live in a time and place where animals infuse our lives. We share our homes with animals, and make ourselves mentally and physically healthier by doing so. We vacation in national parks and other animal-rich areas because we want to witness our companion animals' wild counterparts and their behavior. Most of us eat animals and dress ourselves with animals. The sports teams we root for take on animals as symbols, and the cereals we buy are sold to us by talking animals. The tales we read to our children are inhabited by animals who impart wisdom; the poems, novels, and adventure stories we read may ignite our felt connection to nature.

But *why* should it be this way? Why are we humans emotionally invested in, and sometimes transformed by, close encounters with animals? Why does a multimillion-dollar pet industry in the United States thrive even in tough economic times? Why do most major cities invest in a zoological park and an aquarium? Why are sports teams—not only the Lions, the Tigers, and the Bears, but also the Jayhawks, the Mud Hens, and the Marlins—so often named for animals? Why are the television shows *Animal Planet* and *Nature* so enduringly popular? Why is an animated mouse named Mickey recognized instantly around the world? Why do our emotions at being with animals so often spill over into religious experience? In *Being with Animals*, we will journey through prehistory and history, and across the globe in the present day as well as the past, in order to answer these compelling questions. As we go, we will bring *animals, emotion, and evolution* together with *religiosity*, humans' expression of religious awe. We will look over the shoulders of scientists who offer the latest insights from anthropology, archaeology, and studies of mammals and birds.

For me it's a natural, bringing together these four threads. I'm an animal lover, and have been since childhood. My personal life revolves around family, which we define to include a rather stunning number of domestic cats (and the occasional rabbit). My professional life as an anthropologist is chock full of monkeys and apes; I lived for fourteen months in Kenya in order to track and observe baboons, and more recently have observed gorilla and bonobo groups closer to home. I teach and write about evolutionary matters, with a focus on humans as the apes who became upright walkers, speech talkers, and believers in the supernatural.

I believe that one of our most profound connections with animals lies in our emotional experiences of the world and each other. For one animal, *Homo sapiens,* that world of emotion became a world of religious ritual, and it fascinates me how that happened and what part other animals played as it came about.

And I love to tell animal stories. The stories offered in this book are grounded in three themes. Animals—the mammals and birds of which I write—are complicated beings. They may bond as friends, and offer to each other comfort in times of trouble. They may also snarl, snap, attack, maim, and hurt each other, even outside of the predator-prey context when they are just hunting for a good meal. No animals are "noble savages," expressing a simple ethic of goodness and simple compassion, and no animals are only efficient killing machines at work in a nature run red with blood. Like us, animals tend to express complex facets of their being. Like us, they have personalities and moods, and animal equivalents of grumpy mornings and sunny afternoons. All this is as true for animals when they go about their daily lives with each other as when they interact with us.

This complexity comes about because of variation. Even within a species, individual animals differ dramatically in how they tend to respond to events. Some of those differences probably stem from animals' genetic makeup. Genes matter! Genes may cause an animal to tend to be more shy, or more gregarious. Still, a resounding principle to emerge from recent science studies is that animals are highly responsive to the ways that they are reared, and to their immediate surroundings—just as we humans are.

Of course, we humans *are* animals. This is my second theme. It's only for simplicity's sake that I write as if *humans* and *animals* are somehow separate categories. The mutual relating we engage in with other animals transforms us, yes, but that transformation rests squarely in the common evolutionary trajectory that we share with other creatures. We, all of us, have evolved, and changed over time. *Homo sapiens,* those evolutionary newcomers in a scene choked with animals and plants in thriving prehistoric ecosystems in Africa, evolved to think and feel with *other* animals right from the start. To explore being with animals is to explore our own past, from hunting and gathering lifeways on the African savannas through Ice Age art through early settled villages in Anatolia and beyond.

And finally, being with animals may heighten our senses through a renewed appreciation of the beauty of evolutionary continuity or a deepened sense of God's hand at work in the world, or both. We find, indeed we create, our best selves through animals because it is only other animals who can offer us the transcendent experience of shared ways of being in the world.

This transcendence sits, ironically, side by side with the reality that some humans, in some postindustrial, frenzied, all-about-ourselves societies, have lost the knowledge—and the feeling, the visceral certainty that lives in the body as well as in the brain—that we are part of a community of animals. Yet *lost* cannot be the right word, because the knowledge and the certainty is there: hidden perhaps, but there nonetheless. It's just as the writer Thomas Berry says: "[Animals] provide an emotional intimacy so unique that it can come to us from no other source. The animals can do for us, in both the physical and the spiritual orders, what we cannot do for ourselves or for each other."[6]

Animals and emotional intimacy—and evolution. Let's start there.

2

Humans Emerging: From Savanna to Art Cave

● ● ● ●

There is a wolf in me . . .

I keep this wolf because the wilderness gave it to me and
the wilderness will not let it go.

—Carl Sandburg, "Wilderness," 1918

GALLERY AFTER GALLERY, the art dazzles. In one massive
frieze, lions are depicted in profile. The strong lines of the
tawny bodies, the intensity of the gaze, and the tautness in the neck
muscles all combine to convey *predator*. Bison, rhinos, and mammoths
are here also, as is a mysterious figure with the heavy head of a bison
melded with the lower body of a woman.

On close inspection, the subtly masterful aspects of these images
emerge to our vision. An exhibit catalog of sorts describes them: "One
of the lions seemed to dominate the pride with its penetrating gaze;
its head was drawn with darker lines. With theatrical effect, a horse
had been placed at the back of a niche and some bison heads were
superimposed like hunting trophies. A rhinoceros was endowed with
multiple horns that suggested movement."

In a nearby room is another large panel. Here it's the horses that immediately draw the eye. Three horses are depicted with slightly bowed heads and peaceful expressions; a fourth, boldly darker in color than the others, stands serenely looking forward. Other galleries reveal a diverse zoo of images that suggests an intimate familiarity with the animal world.

These glorious images are found in none of the world's famed art museums, but in an ancient French cave. The "exhibit catalog" is the beautiful book *Dawn of Art: The Chauvet Cave, the Oldest Known Paintings in the World,* and as its title suggests, the single most stunning fact about this art is its age.[1] The images are about 30,000 years old. There's no more obvious place to launch our evolutionary exploration, for these images offer thrilling clues to the anthropology of early humans and their relationship with animals.

Chauvet Cave lies in a gorge-filled region of France called Ardeche, located between Lyons to the north and Marseille on the southern coast. A trio of spelunkers—Jean-Marie Chauvet, Eliette Brunel Deschamps, and Christian Hillaire—had discovered other caves before, but nothing in their previous experience prepared them for what happened when they stumbled onto Chauvet one day in 1994. Walking along an ancient mule path beside the tall cliffs, they noticed a small cavity two meters above the path. They entered via a small vestibule, crawled through a narrow passage, and (after a return to their car for the appropriate equipment) descended to the cave floor in darkness. Eventually they reached the art galleries, where they reveled in image after unexpected image:

> During those moments, there were only shouts and exclamations; the emotion that gripped us made us incapable of uttering a single word. Alone in that vastness, lit by the feeble beam of our lamps, we were seized by a strange feeling. Everything was so beautiful, so fresh, almost too much so. Time was abolished, as if the tens of thousands of years that separated us from the

producers of these paintings no longer existed. It seemed as if they had just created these masterpieces. Suddenly we felt like intruders. Deeply impressed, we were weighed down by the feeling that we were not alone; the artists' souls and spirits surrounded us.[2]

Souls and spirits. What clues can the images in Chauvet Cave, and sister caves throughout Ice Age Europe, like Lascaux in France and Altamira in Spain, give us to the nature of ancient animal-human relating? Does the "being with animals" of evolutionary times include any religious overtones?

Homo sapiens evolved in Africa 200,000 years ago from humanlike ancestors, who in turn had evolved from apelike ancestors. A central plot point in the story of our own evolutionary trajectory is our shared, if edgy, coexistence with other animals. This narrative gives us a "backstory" to the glorious explosion of animal-oriented art at Chauvet Cave.

BEYOND THE HUNT: HUMANS EVOLVING

As I write these pages, Lucy is visiting the United States from Ethiopia. Lucy is a fossil australopithecine, named by her discoverers after the Beatles song "Lucy in the Sky with Diamonds" that played in their camp during the 1974 archaeology season. Wherever she appears on her Stateside museum tour, people get a chance to stand mere inches from a touchstone in our own evolving story.

A little over 3 million years ago, Lucy lived with others of her kind (*Australopithecus afarensis*) at a place called Hadar in Ethiopia. Though her brain was only the size of an ape's, she strode around as no ape can do for very long—bipedally, or upright on two feet. Scientists have pored over Lucy's bones and teeth, as well as those of other australopithecines. The bones tell us a great deal about what Lucy ate, how she moved, and even that she probably made tools like a modern-day chimpanzee's. We know next to nothing, by contrast, about what kind of groups she lived in—but like other primates, she must have depended heavily on others in a close-knit community to survive. Though small-bodied (Lucy stood under four feet tall), and lacking any but the most makeshift weapons, australopithecines lived in ecosystems fairly stuffed with fierce predators.

It takes no Sherlock Holmes to deduce that these early ancestors of ours were more humble prey than fierce predator.

Marks on skulls and bones show that leopards, saber-toothed cats, and hyenas often made a meal of the australopithecines. The famous "Taung child" found by the South African anatomist Raymond Dart in 1924 was the victim of a predator. The child's skull bears the clear marks of an eagle's beak and talons. Judging by extension from what we know about the intensity of the mother-infant bond in apes, it must have been a fearsome event to witness: an eagle swooping down on an australopithecine family, then snatching away its most vulnerable member.

The Taung child's fate, pinpointed as to cause only in 1995 by scientists Lee Berger and Ron Clarke, is an ironic one.[3] Dart himself was convinced—and famous for insisting—that australopithecines were fierce and bloodthirsty killers who thrived on bringing down animals for their meat. For years, anthropologists pledged allegiance to a single dominant theory: Man the Hunter. It was the hunting of wily, clever, and dangerous big game that pushed early people to cooperate in groups, and early brains to enlarge, according to this theory. Only slowly did scientists uncover enough fossil evidence to realize that hunting was not a dramatic or driving factor in our lineage's becoming smart and social (smarter and more social than any ape).[4]

As the archaeologist Steven Mithen notes, human-animal interactions began to change after about two million years ago.[5] Three related factors combined to vault our ancestors into a role of increasing dominance in the landscape: bigger brains, brawnier bodies, and better tools. The prehumans living at that time, classified in our own *Homo* genus to highlight their differences from australopithecines, butchered carcasses with stone tools and cracked open animal bones to extract fat and marrow.

Yet this technology-aided taste for meat did not translate into a pre-occupation with hunting at this point in our story. It may be less noble a heritage, but all clues point to Man the Scavenger rather than Man the Hunter. Mithen paints a picture of "largely sneaky scavengers, creeping in after the lions, hyenas, and vultures had had their fill."[6] In fact, these male-oriented phrases (Man the _____, fill in the currently hot buzz word) are pithy and memorable, but they make for poor science. Based on what is known from modern hunter-gatherer groups, anthropologists project into our past a significant role for women in providing food for their groups, through the harvesting of vegetables, nuts, and fruits. These foods were ready, steady, available staples that acted as insurance buffers when meat, fat, and marrow were hard to come by. For years, scientists have bickered about "Man the Hunter" versus "Woman the Gatherer" theories. Far better to acknowledge the key contributions of scavenging and gathering in our prehistory, and to admit that we have no clear idea which sex did what, that far back in the past. (In some modern hunter-gatherer groups, women do hunt.)

Despite scientists' certainty about scavenging and gathering in the early *Homo* era, few other subjects continue to be as mythologized as our big-game-hunting origins. Elizabeth Marshall Thomas, a sensitive observer of hunter-gatherer peoples' interactions with animals, nonetheless falls prey to this myth when she writes:

> Our species seems to have an atavistic urge to hunt such as is found among the carnivores. That our impulse is atavistic can be seen in the reaction shown by some of us—not all of us, surely, as we have been overfed and sedentary much too long, but some of us—when, say, we notice a deer at the roadside. . . . Physically speaking, we are ready to run forward. No matter how often this happens, our reaction is always the same and does not fade with time or repetition.[7]

Now, I'm not sure with whom Thomas spends her time! Deer frequent the roadsides (and backyards, in some cases) in Virginia where I live, yet I know few people who must restrain themselves from rushing forward toward the animal, in an atavistic urge to spear it. True, our closest living relatives, the chimpanzees, hunt monkeys as a matter of course, and become highly excited when doing so. Thomas describes this emotional arousal in startling terms: "Red meat was hot, and the bodies of the monkeys were active and struggling. When you grabbed them, they would open their mouths very wide and their pupils would grow huge and black. . . . You could smell their fear and feel their hearts pounding."[8]

Perhaps chimpanzees do thrill to their prey's fear. Yet, in this context, chimpanzees are no guide to the behavior of our early ancestors. Chimpanzees do not scavenge and early *Homo* did not hunt. Even if Thomas's "atavism" refers narrowly to our own species only, I still object, simply because the gathering of fruits, vegetables, and nuts was far more critical at that stage. In fact, I began to wonder why writers besotted with our past don't rhapsodize about the atavistic lure of fruits and vegetables! (As so often happens with these things, mere days after I first had this thought, I discovered a book at least somewhat related to it, called *The Fruit Hunters: A Story of Nature, Adventure, Commerce and Obsession.*)[9]

When does hunting begin? A mix of bones and tools at Boxgrove in Sussex, England, makes a convincing, if not airtight, case for hunting back to 500,000 years ago. At Boxgrove, bones of rhinoceros, horse, red deer, and other species are mixed with crafted tools of flint. Cut marks on the tools are primary—that is, laid down before any teethmarks from other scavengers. The animals were butchered by Boxgrove dwellers. How they brought down hulking rhinos is unclear. A single tantalizing clue comes from a horse's ancient shoulder bone, pierced by a projectile.

Three ancient wooden spears usher in the age of undisputed big-game hunting. Dated to 400,000 years ago, these were found at Schöningen, Germany, and announced by archaeologist Hartmut Thieme in *Nature*: "Found in association with stone tools and the butchered remains of more than ten horses, the spears strongly suggest that systematic hunting, involving foresight, planning and the use of appropriate technology, was part of the behavioural repertoire of pre-modern hominids."[10] Prehuman, but skilled, horse hunters: from this point on, innovations in weapons and strategy propelled the craft of hunting forward.

Big-game hunting, then, is the second great change in the trajectory of animal-human relating. Now there was an intense pressure on our ancestors to track and outwit animals who were themselves well evolved and well adapted to the local environment. Those prey animals were clever creatures, tuned in to the dynamic seasons and their shifting resources—and, unlike the Schöningen speared horses, some of them were adapted to eating (or trampling or goring) the humans who shared their turf.

The enduring dangers of hunting coupled with big payoffs. What an abundance of resources came along with these edible packages of protein and fat! Hides could sheath a cold human body; mammoth tusks could support the building of shelters. The resourceful use of animals and their parts was on a skyrocketing upward trajectory that would continue into the modern era (consider the way that American Indians of the Plains historically used every part of the buffalo, from skin to internal organs).

Few of us living in modern cities, towns, or farms reel from close encounters with dangerous animals. Recently, driving home on a spring day along the York River in Virginia, I came across a large snapping turtle endangering its reproductive success by trying to cross to the opposite woods, headlong through traffic. I thought to rescue it with a quick carry across the road, but perhaps I resembled a hungry predator, for the turtle began to lash its head around and lunge with its beak toward me. (I herded it, hands-off, to safety instead.) Also, a feral cat under my care gets ornery now and again, and lashes out with claws and canines. Not since a visit to Yellowstone National Park, however, with its large populations of bear and bison, have I had any reason to fear for my life and health because of sizable or aggressive animals.

It's no good, though, to oversell the notion of human dominance in these ancient ecosystems—or even in modern ones. For millions of people, it's no casual matter to coexist with dangerous animals. As Donna Hart and Robert Sussman report, between 1975 and 1985, tigers killed 612 people in one large, forested delta region in India and Bangladesh (at the Ganges and Brahmaputra Rivers). On the very day I am writing, CNN is reporting that a wild elephant has killed seven people and injured twenty-four more in a village in northern India.[11] (When the elephant began to destroy crops, the villagers surrounded her, and began to scream and bang drums at her; the animal was clearly terrified and eventually was killed.) These examples show that while the *pattern* of animal-human relating has changed drastically since our prehistoric past, our species remains under pressure to deal with predators via brain-and-technology-based ingenuity. Even a low-tech solution seems to make a positive difference in tiger-infested India: when people affix face masks to the backs of their heads, tigers attack less often. The eyes on the mask seem to convey a message of alertness, and the big cats choose to hunt elsewhere.[12]

As important as all the survival-and-feeding adaptations were for our prehistory (and our present), what's most crucial for our story of animal-human relations reaches into a more soulful realm, one where utilitarian aspects are coupled with, or even crowded out by, aesthetics.

THE COLORFUL PATH TO CHAUVET

Color seems to have captured our ancestors' fancy, so that they used reds and pinks in special ways. A faint hint of this practice appears at the site of Atapuerca in Spain, where twenty-seven humanlike skeletons, dated to 350,000 years ago, crowd the bottom of a deep pit. Mixed among the bones is a hand ax crafted from quartzite. Such tools were in common use at this period, well after the invention of spears. Used by a pre-*sapiens* ancestor, they aided in the butchering of animal carcasses. The Atapuerca tool, though, is startling in its color—pink!

It is the *single* human-made object to be found together with the skeletons.[13] Was this tool a grave offering? Might it point to deliberate interment of the bodies, or even some kind of funeral ritual? Did the pink color symbolize something to the Atapuerca people? This particular association of color, tool, and bones stands alone for its time period, so all we can do is invoke the old standby conservative clause: use caution in making any interpretation. By around 100,000 years ago, pigments and deliberate burials were linked at *Homo sapiens* sites. At the cave of Qafzeh in Israel, fragments of red ocher stain bones that were buried in an apparent ritual context.[14]

From around the same period, perforated shells of marine gastropods litter sites in Israel and Algeria. Interestingly, the sites themselves were inland. Could these sets of beads represent early symbolic ornamentation or even jewelry? Possibly the holes were made in the shells by people, but they could also have come about through natural processes of wear. What's clear is that the sites' occupants selected shells found some distance from their living quarters, either seeking those with perforations already present or creating the holes with an eye for a certain look that meant something special to their group.[15]

A site along the Indian Ocean in South Africa affords us the richest record yet available of early aesthetics. Seventy-five thousand years ago at Blombos Cave, people made a kind of proto-art as distant in time from Chauvet's as Chauvet's is from us today. Blombos people inscribed ocher with geometric patterns in the form of crosshatches and lines. Whatever these marks meant—and we really have no way to know—they amount to material culture that flashes to us across the millennia the news that these people could think abstractly.

Most moving to us today, perhaps, is the knowledge that people at Blombos chose to ornament themselves. Ocher adorned the body, as did jewelry fashioned from mollusk shells. Pea-sized shells were perforated and strung together in what may have been necklaces or bracelets. At Blombos, unlike at the Israel and Algeria locations, the mark of the human brain and hand on the jewelry is quite clear: these were created, not merely found. The shells' wear pattern suggests, too, that some type of leather or twine was used to string the items together.[16]

A marine-rich site, Blombos allowed its inhabitants to mine the sea not only for survival but also to make symbolic statements about their own appearance and identity. What we see in the material culture does not quite amount to a statement about animal-human relating. I cannot convince myself that the Blombos people enjoyed some kind of strong connection with the mollusks of their world. Yet, from Blombos, we learn another way in which the animal world is brought into the human world in creative ways.

We can only wait to discover what else may be learned about the creativity of early *Homo sapiens*, now that scholars are seeking evidence for symboling even before the time period of the Ice Age art caves. And it's not our species alone that surprises; our cousins the Neanderthals do, too. Everyone knows something about these shambling, stuttering, thick-headed, cave-dwelling mammoth hunters. But not so fast with the stereotypes! More and more traces of an aesthetic sensibility are turning up at Neanderthal archaeological sites, forcing a quiet revolution in scientists' view of these creatures. In terms of art, the pièce de résistance is the object dubbed the French Neanderthal mask.

Around 35,000 years ago, Neanderthals took a triangular piece of flint, modified its shape somewhat, and drove a piece of wood through two openings; the result is an object that looks strikingly like a face. Dated to just about the same time period as Chauvet, the French mask is a fascinating symbol in two quite different senses. It was a symbol for the Neanderthals themselves—their choice to create a human face suggests a reflective capacity for matters of appearance or identity. But it symbolizes something, too, for a comparative look at Neanderthals and early modern human beings. As advanced as the creative act behind the mask is, it cannot hold a candle to what the Chauvet artists were just about to do. Indeed, the Chauvet people were probably representative of the *Homo sapiens* groups who, soon enough, outcompeted the Neanderthals. By about 27,000 years ago, the Neanderthals had vanished. From then on, the *Homo sapiens* line-age was alone in its evolutionary ascendance.[17]

Why the Neanderthals disappeared is still a conundrum, not least because these people were skilled hunters of cold-adapted prey. There's even reason to think that in some regions Neanderthals had started down a path of symbolic engagement with animals. Two sites in France, from different time periods, yield key clues. At Grotte du Renne, around 33,000 years ago, Neanderthals modified the teeth of bear, wolf, and deer, and wore them as pendants. At Regourdou, around 65,000 years ago, Neanderthals incorporated bear bones into the deliberate burial of one of their own, then marked the grave with fire and an elk's antler. Nearby, more bear bones were very carefully arranged, and some anthropologists think bear meat was also consumed at a kind of funerary ritual. As usual, we're left without a clear grasp of what the animals meant—we'll see the same even at Chauvet Cave—but through the centuries what comes down to us is a sense of the *power* of animals to matter in our past.[18]

Clearly, it's a mistake to claim, as some textbooks still do, that the decorated caves of Western Europe represent a sudden revolution of symbolic thinking, a trajectory that springs to life without antecedent roots. On the contrary, the symbolic life evolved gradually, and animal-human relating right along with it. Straight back to Lucy's time in Ethiopia, and even before (when symbolic behavior was, at the least, invisible to us, and quite possibly absent altogether), our ancestors must have been rewarded for keen watchfulness toward animals in the way that mattered most—increased survival and passing along of their genes to the next generation. *Animals mattered always.* They mattered first as fierce takers of life, later as prodigious sources of nutrition, clothing, and shelter, and finally as those things still, but also as forces in our world that symbolize strength and power, life and death.

The ecologist Paul Shepard, in his book *The Others*, elegantly conveys how acute our imagination must become in order to grasp fully the role of animals, "the others," in our past: "Our species and our best observers emerged in watching the Others, participating in their world by eating and being eaten by them, suffering them as parasites, wearing their feathers and skins, making tools of their bones and antlers, and communicating their significance by dancing, sculpting, performing, imaging, narrating, and thinking them."[19] What a pithy review of the various ways animal mattered in our past! With his concluding focus on dance and art, Shepard invites us to consider Chauvet and other ancient cave art in greater depth.

Our evolutionary story *can* be told primarily as a series of milestones: when we first made stone tools, captured fire, speared big game, and uttered the first words (and made beautiful art). But to do it justice, we need to look at Chauvet and other sites to grasp how deep ran our emotional connection to other animals and to the larger world around us, as well. It is helpful to hold in mind the record of gradual, ever more complex, ever more symbolic relating with animals, while at the same time recognizing that there *is* something wholly novel about the time period of the Ice Age caves. Now, art blossoms into its fullness, and, with it, evidence for new depths of animal-human relating.

CHAUVET, THE BEAR CAVE

Chauvet was a bear cave.[20] Visible traces of bears are found everywhere within it today: bear bones, skulls, and teeth are the physical remnants, but also found are claw marks and prints, reminders of moving animals who lived their everyday lives in the cave. Visitors can even see precisely where on the cave floors bears once slept; circular hollows, some marked by claws or fur, still remain visible. The incredible art images that grace the cave walls blazed to life in a space inhabited by both humans and bears.

Did the two species meet and interact? Was some kind of bear cult headquartered at Chauvet? In order to probe the nature of human-bear relating, we need to step back and look first at what traces of Ice

Age people themselves emerge from the cave's depths. A second marvelous book, *Return to Chauvet Cave*,[21] is an ideal resource.

As best we can tell, the cave was occupied by *Homo sapiens* at two distinct time periods. Starting 32,000 years ago, the cave became a veritable and vibrant artists' studio; people brought in materials for lighting the cave, and for making the art. In a burst of incredible activity, the glories of Chauvet were created. Two thousand years later, people abandoned the cave; now its sole occupants were bears. People once again used the cave in some fashion around 26,000 years ago. No art was created at this time, as far as we can tell, but today's cave trekkers can still thrill to the footprints of this second wave; a preserved 230-foot-long trackway evokes the path of a young boy as he wandered through the cave's chambers.

Evocative, too, are the human handprints that mark the cave. Over and over, Chauvet people pressed their ocher-colored palms against the walls. How fitting! The hand is at once a uniquely human appendage that allows us to create images with tools, and also the quintessential mark of a primate and thus a symbol of our long, shared evolutionary history with other animals. One man (it's usually possible to "sex" a handprint, so we have a pretty good idea the person was male) had a curved fifth finger on his right hand, a trait that became his signature of sorts. Through this material signature of his humanity, we glimpse at Chauvet a small marker of a life that embraced beauty as well as hardship.

If individual difference is written into the archaeological record at Chauvet, so is a kind of cultural unity. Recall that I suggested a "burst" of incredible art activity in a short time frame. But how can we know that? Couldn't the paintings have been created instead at a slower rate, perhaps a few per generation over the 2,000-year occupation? The prehistoric-art expert Jean Clottes concludes, in *Return to Chauvet Cave*, that "most figures were drawn by a very small number of people who shared the same ideas, probably during a fairly short length of time, even during the period when the cave was first visited by humans."[22] The painters routinely incorporated topographic features of the cave into their images, as when a rock's fissure is used to suggest an animal horn, and they employed certain styles in their depictions, as when the ears and legs of rhinos were drawn in precise ways over and over again.

Patterns such as these do point to shared ideas among Chauvet's artists. Probably the styles are conventions, the product of shared discussions rather than of independent choices made by artists working in isolation from each other. After all, Chauvet people had language; they would have discussed not only where to find game and how to bury their dead, but also what and how they painted. This cultural sophistication raises its own questions. Could the older, more experienced Chauvet artists have passed along their techniques to apprentices? Could the artistic traditions visible at Chauvet have been conserved over time? I think the jury is still out on the question of how many people painted at Chauvet and in how compressed a period of time.

Also, the images aren't as uniform as Clottes's comment might suggest. In the horse images at Chauvet, for example, no patterns exist. Some horses are drawn in simple profile while others are splendidly rendered. Only half are complete, full-bodied animals. Some are drawn in black, others engraved, and so on; no conventions were used akin to the distinct and repeating rhino ear-and-leg complex.

Mysteries still abound, then, about Chauvet. For our purposes, the important part is not so much the who, when, and how of the cave art, but the what and the why. What species are represented, and why these? Why did people paint animals in such overwhelming abundance, yet choose to record their own images hardly at all? Fourteen animal species are found in Chauvet's art. Of all Ice Age caves, only at Chauvet are found an owl image and a panther image. I find the owl quite interesting. Not painted or drawn, but instead traced by fingers on a soft wall, the picture is unmistakably an owl. Yet it is one of the very few images at Chauvet that strikes a viewer as simple, even childlike, in its rendering.

More significant for the *what* is the atypical distribution of predators versus prey in the images, set against the ratio from other decorated caves. Rhinoceroses, lions, bears, and mammoths—the largest and most dangerous animals—outnumber the smaller species, those that would have been food (prey) animals for Chauvet people. Yet there's no sense of danger or human-directed menace in the images. In one fascinating scene, two rhinos face each other—indeed, are almost touching, head-on. Are these two about to begin a conflict? Did the artists wish to render a scene that they had witnessed? Why so few scenes like this, in favor of isolated animals or clusters where the individuals do not engage? The cave walls remain silent on these questions, but it's safe to say that the art is about a lot more than hunting.

Indeed, the Chauvet galleries blaze with a message: Our ancestors' consciousness was filled with animals, not just as food, or as danger, but as something more. And this brings us right back to the bears.

Scientists have worked out that bears lived in Chauvet before, during, and after the time periods of human occupation. In some cases, bear claw marks are overlaid on top of the art images, whereas in others, it's the claw marks that were clearly laid down first. Dating of the Chauvet bones reveals that people and bears visited the cave during the same period.

Bear bones! We are talking about impressive numbers: over 170 bear skulls and about 2,500 bear bones clutter the floor, sometimes in groups and sometimes singly. In one chamber, Chauvet people placed a bear skull on rock that had fallen onto the floor. The anthropologist Joelle Robert-Lamblin notes that scattered around this feature, "over a radius of 7 m (23 feet)," are "about forty other bear skulls (mostly accompanied by the rest of the skeleton). . . . This very confined place, where numerous bears died, could have been the inspiration that caused the cave's human visitors to choose this area for the celebration of a bear 'cult.'"

Here again, I urge caution. To wring specific meaning from the arrangement of bear skulls in Chauvet is tempting, but a leap to the conclusion of a bear cult is too hasty. Robert-Lamblin goes further than I would in her interpretations of a bear cult, yet in two ways she's on to something intriguing. In pointing out that at Chauvet "the bear seems to be entrusted with an eminent role," she's underscoring the symbolic role played by bears at the site, a role that exceeds that predator-prey duality that for so long characterized our ancestors' relationship with animals. *Something else is happening here.* The bear is neither only a thing to be feared nor only a thing to eat: it's symbolizing something more for Chauvet people (as perhaps it did at Regourdou for our Neanderthal cousins).

Further, when she wonders if the bear was perceived as an "intermediary between men and animals," Robert-Lamblin leads us right into fascinating territory—the territory of the sacred. An expert on the anthropology of circumpolar peoples, Robert-Lamblin explores analogies between Arctic cultures, ancient and modern, and Chauvet culture. Among the Inuit (sometimes called the Eskimo), the bear is often featured in rituals, and is seen as a go-between for humans and spirits. Here, of course, is a potential connection with the sacred, if indeed Chauvet people conceived of something like a sacred arena. Like many others, as we will see, Robert-Lamblin emphasizes the "dark and fearsome subterranean world" of Chauvet, a place "probably inhabited by spirits in the eyes of the humans who ventured into it with their limited means of lighting."[23] Is she right? Would Chauvet be a place ripe for human participation in an otherworldly realm where animals and humans are linked by spirits?

The question is provocative. Specific analogies between two wholly different cultures—Inuit and Chauvet—raise high my anthropologist's red flag. The Inuit worldview, according to Robert-Lamblin, imbues all living beings with a soul. Rituals and symbols help negotiate that edgy coexistence between humans and other animals— brother species in many ways, with a mythical shared past and an ever-entwined present. Yet I wouldn't want to map these aspects of Inuit cosmology onto Chauvet's culture. While it's true that the Chauvet people's world may have resembled the Arctic peoples' in some specific ways (given the Ice Age environment), we need more concrete evidence than just the placement of bear skulls in the cave and an analogy with another culture to shore up a claim for animal-based religiosity at Chauvet. Do any of the Chauvet art images themselves hint at religiosity?

In a fraction of ancient images, humans *become* animals or animals *become* human; that is, the figure represented is some kind of an amalgamation or hybrid. These require a closer look.

ANIMAL-HUMAN TRANSFORMATIONS: CHAUVET AND BEYOND

It's one of Chauvet's most arresting images: a black figure, with the substantial fleshy legs, broad hips, and unmistakable pubic area of a human woman, topped with the head and chest of a bison. Flowing from the bison is a human arm that ends in a hand with long fingers. Parts of the figure seem to have been deliberately rubbed out, and the wall on which it was created has eroded; what is visible today may not be the original image. Even so, that a composite animal-human figure was intended is clear.

To Robert-Lamblin's eye, other images at Chauvet are anthropomorphized. Not full composites like the bison-woman, these pictures of animals hint at humanity. Some lions are drawn in ways "strongly evocative of human profiles." Images of the human hand occur in association with animals. Sometimes it's a matter of mere proximity, but other times, the handprints contribute to the image itself. In one case, ninety-two palm impressions, made up of dots of ocher, create the silhouette of a horse.[24]

At France's most famous art cave, Lascaux, a curious male human figure (courtesy of its prominent erection, the person is unmistakably male) was given the head of a large-beaked bird. Drawn quite primitively, with that one notable and tumescent exception, the figure sports a long, featureless torso and stick legs. Next to this figure on one side is a bird perched on a pole, and on another, a wounded bison and rhinoceros.

At Fumane Cave in Italy, where if anything the images appear to be slightly older than those at Chauvet, there's an intriguing slab of rock, fallen from the cave roof, that shows some kind of hybrid human-animal figure.

Of course, not all ancient art is painted, drawn, or traced on cave walls. Upon opening the thick and spectacularly illustrated book *30,000 Years of Art,* one finds that its first entry is a sculpture carved from a mammoth tusk.[25] Called the Lion Man of Hohlenstein-Stadel Cave, the figure, slightly stooped because it follows the natural curvature of the tusk shape, has a feline head, whereas the body looks mostly human. But its humanity is blurred and ambiguous, because the limbs do not end in clear hands or feet, but just taper off. As its primary position in this millennia-spanning book indicates, the Lion Man was made in Germany at the same time as the Chauvet art, about 30,000 years ago.

Recall that human figures generally are rare in cave art. By about 13,000 years ago, in Addaura Cave in Italy, we find a limestone engraving that depicts a number of humans together. Unlike the passive, free-floating nature of most of cave art's animal images, here the scene is of some kind of group activity. In its "degree of dynamism," this engraving is groundbreaking.[26] Some figures walk or stretch their bodies while six people circle around two others. What is going on? Is this a dance or a ritual of some sort? It appears so, but assignment of specific meaning is problematic. Notably, animals are engraved around the periphery of these same walls.

Joining the ubiquitous animal images and the far rarer human images, then, are animal-human hybrid images from before 10,000 years ago. To describe them is easy enough, but as I have already noted, to interpret them is another matter. How can we think about these images? The archaeologist Randall White, musing on the Fumane art and others of its ilk, noted, "It is one thing to represent a horse, but another thing to represent something that is a figment of the collective imagination, something that doesn't exist in reality. People had ideas about the world that were abstractions, which we can only describe as religious."[27]

Is a religious interpretation warranted for the animal-human hybrid figures? Some analysts have been quick to suggest that these images, especially the bird-headed man at Lascaux, link somehow to shamanic activity. Shamans are religious figures who, in many cultures today, relate intimately to animals and may even transform themselves into animals as they perform rituals of healing or contacting the ancestors (see chapter 6). Others find this same idea to be a preposterous leap.

The archaeologist Paul Bahn's derision for a projection of shamanism into the past itself leaps off the page. He notes that the art hybrid figures could just as easily be imaginary creatures as anything else. Randall White links the prehistoric imagination with religion, but need it be so? Could animal-human figures amount to free-flight imagination about the animal world, and its relationship to the human world, shorn of any religious overtones? Does the invocation of a mythical creature like the unicorn, let's say in medieval art (chapter 4), necessarily link to the religious?

To consider these questions, we need to broaden back out from the subset of hybrid figures to the art images more generally. Looking at the collective of art inside the French, Spanish, and Italian caves, we find that animals constitute the major theme. Paul Bahn notes "the overwhelming overall dominance of the horse and bison among the Ice Age depictions."[28] These species, together with rhinos, mammoths, and the like, are rendered naturalistically in one sense, as we have seen: their anatomy and, even more, their manner of being, their ferocity or serenity, comes home to us across the millennia. Yet, most often, these single creatures are divorced from any kind of natural context. When more than one animal is depicted, there's almost never any background, whether a landscape or other animals. There's rarely even any hint of one animal engaged with another. Among the fascinating exceptions are the face-to-face rhinos at Chauvet, two face-to-face mammoths at Rouffignac Cave, and a reindeer at Font de Gaume Cave in France that bends forward to lick a female reindeer kneeling nearby.

What do the images mean? We're back to the why question, the timeless question, the question that gets asked over and over. Are the images' meanings unknowable to us because our ancestors' minds are impenetrable to us? Or does the process of careful analysis and analogy, one that we have already begun, help us grasp something useful in the realm of meaning-making?[29]

If we allow history to guide us, caution in theorizing is indicated. One early theory insisted that the cave walls exploded with expressions of intense human sexuality (because many images were seen to symbolize the vulva and the penis). This idea fails to hold upon close inspection, and has led to a number of chortling comments about the preoccupied mindset of certain early-art scholars. Another theory stemmed from the predominance of prey species on the walls. Did early humans paint the images of animals they wished to hunt in order to boost the hunters' power? This view, called "sympathetic hunting magic," makes some logical sense in the abstract. Yet, as we have seen, it's too simplistic in the concrete. At Chauvet and elsewhere, large and fierce animals were represented, and prey species are only one part of the overall count. This means that the hunting-magic idea can, at best, be only part of the explanation.

As Paul Shepard has invited us to do, pushing past the constricted area of material culture may help us think more insightfully about ancestral patterns of being with animals. As we humans evolved, animals inhabited our bodies (they fueled us with protein and fat) and affected our social lives (they, in part, set our migratory movements), just as we altered their bodies and social lives (through the kill, through the chase, through tracking them). Even the material-filled art caves should stimulate ideas that go beyond the paintings and engraving themselves. Shepard's own speculation dismisses a straightforward "magic" approach in which controlling the animals was the art's major raison d'être. Instead he juxtaposes the images' static nature with the probable dynamic activity of the humans surrounding the images:

"One feels before these astonishing figures in their solitude that their ineffable silence was probably balanced in human song or performance, intended to tell a timeless story. . . . The participants in these ceremonies went inside the body of the earth to attend to whole animals. . . . The telling of the hunt was an adventure in tracking and interpreting signs inside the landscape, in reading interiors of places, an insider's view."[30]

Caves *inside the earth*. Here is one of those phrases that expands our thinking about the past. As Robert-Lamblin noted, the cave is a scary and special place, a dark place illuminated only by the flickering fire-torches carried in by early humans. To us, the caves seem like sanctuaries: amazing places with gorgeous colors and bold lines all around, created by the very people who, in an evolutionary sense, created us. It's impossible to know whether *our* sense of wonder is rightfully projected in the past or, if it is, whether it should be linked to religiosity. Certainly, it's impossible to know *for sure* that the animal images were gazed upon in reverent silence, broken with ceremonial singing or ritual dancing at appropriate moments in some kind of celebration (or appeasement) of sacred forces. But in the end, what's most impossible for me to believe is that our ancestors—who spoke to each other; who buried their dead with elaborate ceremonies; who grappled in some way with the mysteries of life and death, of cataclysmic happenings and everyday events; and who re-created the animal beauty of their world in paintings and engravings of splendid quality—could have entirely lacked a yearning to explore a beyond-the-here-and-now universe.

In Tuc d'Audoubert Cave in France, prehistoric people fashioned bison from clay. Two larger figures, a male following a female, were sculpted against a projecting rock and oversee a smaller, third figure on the cave floor nearby. The animals are gorgeously rendered. Particularly notable is the surrounding context: "In a neighbouring recess, in which survive fifty heel prints of a young human on its floor, pieces of clay had been kneaded into phallic forms. Other blocks of clay lie close to the two bison, ready to be modelled."[31] Tuc d'Audoubert seems to have been home, at about 15,000 years ago, to an artists' clay workshop of sorts. What effort—what time and energy—our ancestors put into their art! Time and effort, we know, often links with ritual; once again we are pushed right to the edge of identifying the roots of ritual religiosity.

Even Paul Bahn, who has long been skeptical of explicitly spiritual interpretations of individual images in the Ice Age caves, thinks it's safe to take that next step and link the caves with sacred matters. The art is often located in deep caves, where, Bahn says, a person experiences total blackness, and possibly claustrophobia, and may lose any sense of direction: "To enter a deep cave is to leave the everyday world and cross a boundary into the unknown—a strange, supernatural world."[32] If we adopt this kind of framework, it's hard to dismiss the animal-human hybrid images, in particular, as evidence for a spiritual dimension to early *Homo sapiens* life.

JOURNEY TO OZ: AUSTRALIAN ART

The European Ice Age caves are magnificent, even magnetic; as we revel in their colors and lines, they draw us into the images and into our own artistic past. Anyone planning travel to France, Spain, or Italy might wish to consult the book *Cave Art: A Guide to the Decorated Ice Age Caves of Europe,* which includes practical information on sightseeing in the caves (or, in some cases, the replicas open to the public), and how to reach them from major European cities. But, just as we know these caves do not signify the origin point for aesthetics in the sweep of human evolution, neither does Western Europe have a geographic lock on early art.

Images of rhinoceros and zebra, created 26,000 years ago: Namibia's Apollo Cave shows that, right from the start, African cave art was animal-focused. Later, the continent's rock art stuns in its sheer variety, and sometimes in its size. In Niger's Sahara region, two

more-than-life-sized giraffes grace a large sandstone outcropping.
Carved about 6,000 years ago, the male is twenty feet long, his female
partner about half that size. The female's gracefully bent neck nearly
touches the male's back; both animals' bodies bear the familiar reticu-
lated patterns that we know today. Each giraffe's mouth emits a wavy
line with a small human figure at its end.

The prehistorian Jean Clottes, who mused about the time frame in
which Chauvet's art was produced, felt overwhelmed at the sight of
the giraffe figures. He describes them as "truly extraordinary for their
size, perfect proportions, and the mastery with which they have been
executed."[33] Like everyone else, he wonders if there's a spiritual aspect
to the giraffes. The linkage of humans and giraffes strikes me as a
pretty explicit marker of the importance of this animal to Saharan
peoples (the artists were possibly Tuaregs, desert-dwelling nomads),
but beyond that we lack enough context to speculate.

For a comparative look at animals in ancient art, a journey to
Australia is particularly apt. Aboriginal art offers unique lessons about
how the past may inform the present.

Controversy surrounds the timing of Australia's colonization,
though the best evidence points to 60,000 years ago. Dating the origins of native Australian art is hotly contested as well. Does the evidence of red ocher (the main pigment used in rock art) soon after
colonization of the continent mean that art was made there so early
on? Probably the answer is no: red ocher is used in burial as well, as we
have seen. If red ocher were the litmus test, the origin of Western
European art would significantly predate Chauvet. A suggested date
of 39,000 years ago for images at a rock shelter called Carpenter's Gap
may or may not hold up, because the evidence is quite scanty.
Meanwhile, the anthropologist Harold Morphy offers a relatively conservative timeline of aboriginal art.[34]

By at least 30,000 years ago, native Australians were making rock
art. Engraved circles and lines, a form of geometric art, constitute
these earliest images, ones we know now were contemporaneous

with the Chauvet paintings. (Was there something about the 30,000-year-old human brain that permitted the flourishing of artistic creativity?[35]) The early geometric theme continues in the first rock paintings, but around 20,000 years ago, animal images are added in, too. From Arnhem Land, a large region of Australia located, today, east of the Northern Territory's capital of Darwin, hunter-gatherers lived in a rather bleak environment. Here, like their Western European counterparts, they brought animals into their art: kangaroos and wallabies, for instance. Only around 10,000 years ago did human figures enter the picture—literally. Sometimes the hunt is depicted; much later, when Europeans arrived, the art teems with ships, wagons, horses, and buffalo.

It's the animal images we're after, in the early art, of course. The *what* and the *why* of these are important, just as they were at Chauvet. But these aspects can only be understood in Australia by plunging into *when* questions, in a most unusual way. To understand what aboriginal art is meant to be about, we have to grasp something of the Dreamtime (or, as it is sometimes called, the Dreaming). Once again, Harold Morphy is a seasoned guide.

Though *Dreamtime* is an English word, it approximates an aboriginal cosmology that centers around a fluidity of past, present, and future. Time is part of it, as its name implies, but so is space: where power is located is as vital to the Dreamtime as is the idea that the world's sacredness has been in place since its ancestral origins, and is ongoing. *Reality* is the part that bears emphasis, for Dreamtime is no fancy metaphor, but a description of the way the sacred world is in the aboriginal outlook, and was, and will be.

Before the time of humans on Earth, the world was created. "Ancestral beings," writes Morphy, "emerged from within the earth and began to give shape to the world." A waterhole or cave entrance arose at the site of emergence itself; the path taken by the ancestral beings soon flowed out as a stream. Just as time is fluid, so is matter: "Frequently, an ancestral being transformed from animal, to human, to inanimate form, swimming like a fish or jumping like a kangaroo, walking like a person, singing songs or performing ceremonies, and transforming into a rock."[36]

The ancestral presence, like the Dreamtime itself, is continuous—while being nonlinear—through the past, present, and future. To look at a certain configuration of rocks in Arnhem Land is to look at a feature made when ancestral kangaroos crouched in the river and were frozen there in time. The rock attests to this past event, but the spiritual presence of the animals lives on. The Dreamtime maps stories directly onto the landscape, and this mapping itself exists in the art. The anthropologist Peggy Reeves Sanday links together the Dreamtime, the landscape, and sacred ritual: "So inscribed, the landscape is filled with messages for human living and well-being. . . . The Dreaming transports people out of themselves beyond the constraints of their individuality by inspiring obeisance to the sacred. The associated ritual complex teaches new generations how to move through the land following sacred signposts."[37]

For those of us raised on a linear concept of time (where what happened years ago stays in the past, and what will happen in the future is beyond our knowledge), the Dreamtime may be a challenge to wrap the mind around, because it exists independently of linear time.[38] The changes and continuities that occur there all run on a track parallel to the everyday events that people experience in their daily lives.

Often the Dreamtime is described as unique to aboriginal Australia, and in some ways it is. But I wonder: Is the Dreamtime wholly different from a more familiar (to most readers of this book) brand of human religiosity? Don't many Christians, Jews, and Muslims feel an active spiritual presence when they visit holy sites in Jerusalem? Doesn't time become eerily fluid when one stands on sacred ground, so that the power of past events carries forward through the millennia, not stopping at the present, but projecting itself onward into forever?

These sets of religious systems are not identical, of course; the three Abrahamic faiths make no claim that Jerusalem and surrounding lands are inhabited by beings who exert spiritual power directly in the present, as is the case in Australia with the ancestral spirits. But perhaps the difference is not so terribly great; that is, not so great as is the universal human tendency to experience time's fluidity when it comes to spiritual matters.

In Western Europe, hints of animal-human transformations come in the form of hybrid images. In Australia, both animals and humans are depicted in ancient art, and oral history tells of ancestral animal-human transformations. How exciting it is to recognize that ancient peoples in a number of places grappled with the mysteries of animal-human relationships. Our feeling today, supported by twenty-first-century evolutionary science, that the human exists in other animals, and those animals exist, too, in us, may have no exact parallel in the past. Yet earlier cultures were clearly fascinated *in some way* with these mysterious connections.

Traditional art practices, including those from aboriginal Australia, may give us fresh clues to how we might think about the meaning of animal images to our ancestors. When analyzing Chauvet's art, the very perspective we took was European: in this view, the art is equivalent to the finished product that adorns the wall or bulges from a rock. How natural this seems to most of us, who have grown up learning about masterpiece paintings and sculptures as the pinnacle of art's offerings to the human spirit.

A few years ago, I waited in the chilly Parisian air of early summer (to the consternation of my teeth-chattering family) in order to enter the Orangerie Museum and view Monet's waterlily series. I wanted to stand inside that vista of sweeping-white walls and stare at the beautiful colors and composition, and feel lost in the blue serenity. People may become passionate about their favorite artists. My friend Nancy Hogg is so taken by the works of the seventeenth-century Dutch painter Jan Vermeer that she has thrown herself into a project that she calls the Vermeer Quest. She's aiming to photograph herself in front of each of the thirty-five known Vermeer works at galleries around the globe. Trips to Dresden, New York, Paris, Vienna, and Washington have her well on her way. Nancy may take an interest in how the paintings were created, but what moves her is what fills her eye and heart: the dangling luminosity in *Girl with a Pearl Earring,* the beauty of blues in *Woman in Blue Reading a Letter.*

Art, in some cultures, is as much about process and participation as about product. As Harold Morphy writes about aboriginal Australia, "Art represented the appearance of ancestral forces in ritual contexts: imminent, transitory, effective in achieving a particular purpose and then discarded, hidden or destroyed."[39] *Transitory art,* like the sanding paintings made by Buddhist monks, who devote days to the process only to destroy the outcome. The monks' message with their art relates to the impermanence of all things, even the good things, in life. In the aboriginal case, the message resides in a merging of past and present. Art connects a person with the ancestral world in ways grounded in the body and in practice, and not just in observation. Once again, Morphy explains:

> Since the design of the painting is itself thought to have arisen out of ancestral activity, and since those living today have their origin in conception spirits from the ancestral dimension, art enables the Aboriginal people to participate directly in that world. Paintings and other manifestations of the ancestral past enable people not just to re-enact those events but to make them part of their present lives and to unite their experience of the world with that of the ancestral beings.[40]

It's as if, through art, a direct conduit opens up to the forces that shape time's fluidity, and shape as well that other fluidity where an ancestor can become an animal, then a human, then a rock. Here we gain a way to think about ancient animal art and its meaning. I do not favor a 1:1 analogy that would impose the Dreamtime upon the Chauvet or Lascaux people, but rather an opening up of our assumptions in order to consider an idea: *making animals* may have been as powerful for prehistoric cave artists as viewing the images once they were made.

When a Chauvet artist created an animal-human composite figure, was he or she transformed by the very act of painting or engraving it? Could that act, carried out in deep cave spaces, have projected the person into a state of heightened openness and connection to sacred forces? Might the same process have occurred when a group of people danced and chanted together as the artist made the image? Or perhaps viewing and discussing the images collectively in dark cave recesses was the central moment. Maybe people entered a sacred space when, in their own specially tailored ritual ways, they used the animal-human images to do what anthropologist Claude Lévi-Strauss called "thinking with" animals. To "think with" animals is to do much more than eat them, or plan how to hunt and eat them. It is to relate with them by thinking how we are like them, and how we are unlike them, our animal brethren.

Part of the reason we, today, are so attracted and attuned to animals is that we are locked together, as we have always been locked together, in a shared journey that spans past, present, and future. It's true that animals "connect us," as Boria Sax writes, "with a history in which people often seemed to live on a grander and more heroic scale than they do today."[41] But they do more than this, too: animals *create* with us our past, present, and future. Like us, they breathe air and thirst for water and food and seek others of their kind, and endure the inevitable suffering that comes with the cycle of life, birth, and death. We humans and animals co-evolve, as we always have since the time of Lucy and the other australopithecines, indeed since the time when there was no human lineage.

From the time of origins of the great animal cave paintings, engravings, and rock art, gradually but inexorably the human lineage began to pull away from all other animal lineages. As our brains expanded and our tongues became loosed, as our skills of survival and artistry and symboling meshed together to create new mental and emotional landscapes, we began to bring animals into our world in increasingly complex and varied ways.

Not too many millennia after the Ice Age art caves, we began to tame the wild animals. And just as the story of animal-human relating throughout human evolution prepares us to predict, the animals shaped us during the processes of domestication as much as we shaped them. This new cross-species intimacy led in turn to deeper layers of thinking (and feeling) with animals, and to symboling with animals in new and varied contexts, both religious and secular.

3

Taming the Wild?

●　　●　　●　　●

*[Wilbur finds Mister Ed, the talking horse, asleep in
his living room]*

WILBUR: Oh no.

MISTER ED: If you had a dog, you'd let him sleep in the
house.

WILBUR: A dog is different. A dog is a household pet.

MISTER ED: Then call me "Rover" and wake me at eight.

—From the TV show *Mr. Ed* (1961–1966)

THIRTY THOUSAND YEARS AGO. People in Africa, Asia,
Australia, and Western Europe discuss at length which animals
to hunt and where it's best to gather fruits, vegetables, nuts, or
resources from the sea. They paint walls and engrave rocks with
images and symbols. They bury their loved ones with sorrow and ceremony. And they do this (or at least many scientists believe they did
so) while enveloped in a sensibility for the sacred, with lives attuned to
a world of spirits, a world seamless with everyday life.

Across the world, in the Americas, no hunting and gathering, no
murmur of human conversation, no art, and no burials occur. North,

Central, and South America are not uninhabited, of course: great
tracts of forests and sweeping plains teem with animals. Species of
seemingly infinite variety walk, climb, creep, or soar among the bison
that plod across the land and the birds that wheel through the sky. But
of *Homo sapiens*, there is only absence.

It takes some skill to stand today among the hurrying pedestrians
and honking taxis in the built canyons of Manhattan or Rio de Janeiro,
or even to drive along the strip of fast-food joints one inevitably finds
in any U.S. town, and summon to one's imagination a world without
ourselves in it. It's a challenge to the human mind borne perhaps of
the human ego: How many of us can envision a humanless world and
not fall into the trap of thinking "nothing's there"? But of course, we
humans are the evolutionary novelty on Earth, and in the Americas
we were an especially late-appearing addition to a world of glorious
flora and fauna.

When people first crossed the Bering Strait from Asia into
America, they opened a new era of human exploration. And scientists
now know that they did not travel alone. Domestic dogs accompanied
the first Americans, in what surely amounted to the very origins of
interspecies communion in this vast New World.

Think of that ancient journey: migrating families, wave after wave
of them organized into small groups, making their way across a land
bridge, dogs at their side. Did the dogs help with the hunt, enabling
the humans to sustain themselves more readily as they traveled? Did
individual dogs "tune in" to individual people, and attend closely to
their moods and signals? Did people and dogs play together? Just how
emotionally intimate was the early dog-human relationship? Science
has no way to answer these questions directly, but the latest data tell
us that between 15,000 and 10,000 years ago, people and dogs began
to bond. That process was never really about "taming," but rather
about changes wrought in each species by the other.

GENES AND BONES

A Holy Grail of archaeological science is to discover the oldest inhabited human site in the Americas. For years, all eyes were on the Clovis Complex, a hunter-gatherer culture (first discovered at Clovis, New Mexico) whose people brought down game with finely crafted projectile points. Based on these "Clovis points," and on other evidence for the hunting of bison and mammoth, the date of 11,000 years ago became the accepted one for the peopling of the Americas. Other clues were waiting in the ground, though; brought to light, they offered some surprises.

The archaeologist Tom Dillehay and his team discovered that 14,000 years ago, at a site called Monte Verde, people built huts and hearths; used mortars, digging sticks, and projectile points to find food; and ingested seaweed in ways that improved their well-being. It's the cleverness of this last behavior that brings these prehistoric humans into sharp focus today. "All nine seaweed species recovered at [site layer] Monte Verde II are excellent sources of iodine, iron, zinc, protein, hormones, and a wide range of trace elements," notes Dillehay. "Secondary beneficial effects of these seaweeds include aiding cholesterol metabolism, increasing the calcium uptake of bones, antibiotic effects, and increasing the body's ability to fight infection. . . . These same species are used today as medicinal plants by local indigenous populations."[1]

If this report causes a "Wow!" response for you as it did for me, hold tight for the biggest surprise of all. Monte Verde is located in Chile, at South America's southernmost tip. If the first Americans reached Chile so early, what route did they take to travel through the Americas? How long did that migration take?

Although Dillehay favors a Pacific coastal route of migration, data on that issue are sparse. To unearth a trail of not only people, but of people and dogs, working their way north to south, would be phenomenal, but so far no one has done this. Recent finds do support a pre-Clovis civilization in the New World. In spring 2008 came the announcement of a 12,300-year-old human occupation of caves in Oregon. The dating of coprolites (ancient feces) revealed this early date. A lack of sophisticated artifacts (akin to the Clovis projectiles or other worked tools) indicates that the caves were just a way station on a migration route, and not a permanent or even semipermanent home.[2]

No dog bones were found at the Oregon cave. No evidence of domesticated dogs shows up at Monte Verde in Chile, either. Still, these sites are important because they establish that the Bering Strait crossing happened before 15,000 years ago, and Monte Verde offers a peek at the cultural sophistication of the earliest Americans. To figure out more about the ancient human-dog relationship in the Americas, we need to turn to dog DNA.

A team of geneticists extracted DNA from the bones of two sets of dogs. The first canine group had lived in Latin America (Mexico, Peru, and Bolivia) in the days before Columbus; the second group lived in Alaska after the start of European colonization. Following a sophisticated "tree of origins" analysis using the ancient DNA from both groups, the team concluded that ancient and modern dogs everywhere in the world share a common origin from Old World gray wolves. Here, then, is genetic evidence for dogs and people crossing the Bering Strait together. "This implies," writes Jennifer Leonard and her colleagues, "that the humans who colonized America 12,000 to 14,000 [years ago] brought multiple lineages of domesticated dogs with them."[3]

No separate dog domestication occurred in the New World, then. It's not that people crossed over the Strait, dogless, then later interacted with wolves in ways that kick-started the process of dogs' becoming domesticated. But then, where are the ancient dogs' bones?

It's not unusual in evolutionary studies to find that the bony evidence doesn't precisely map, timewise, onto the DNA evidence. Not all living creatures' bones fossilize when they die, and it's not always readily apparent where to search for those that do. No known dog bones in North America date as old as the DNA evidence tells us they should—old enough to come from the dogs that accompanied the first Americans. Here is a short list of North American "firsts" that we do know (keeping in mind that, as with all dates, these may be pushed back with new discoveries):

- In Utah, at a place called Danger Cave, domestic dog bones date back to 9,000 or perhaps 10,000 years ago.
- In Illinois, at the Koster site, domestic dogs were buried in graves 8,500 years ago.
- In Idaho, at the Braden site, dogs were buried in direct association with people 6,000 years ago.

These last two sites raise questions familiar to us. In chapter 2, we considered what it meant when people began to bury their dead with care and ceremony: Did burial with grave goods and jewelry signify an an early religious rite? Now the question is, did the interring of dogs with people mean that the dogs played a role in early religiosity? Tantalizing evidence suggests that not only the emotion we feel for our animal companions today, but also our tendency to bring them into our religious lives, has its own evolutionary history.

First things first, however. If the DNA study I have described is correct, the dog populations represented in the Utah, Illinois, and Idaho sites had their origins in the Old World. Dog-related sites in the Old World, then, should be even older; indeed, dog burials show up in Germany and Israel well before any in the New World, at 14,000 and 12,000 years ago respectively. Dog remains in Russia may be as old as 17,000 years.[4]

Where precisely did dogs originate? DNA is of help here as well. For years, scientists suspected that dogs originated in the hotbed region for domestication, the Middle East. A nifty analysis shows otherwise. The procedure goes like this: Take representatives of modern dogs from all across the world, 654 of them. Confirm genetically what's long been known, that all these dogs descended from wild wolves. Compare the variation in the DNA across regions to find out where the most genetic variation occurs (a key step for one reason: the variations need time to accumulate, and so "the greatest variation" amounts to the same thing as "the oldest"). Calculate the results, and conclude an origin of the domestic dog in East Asia at about 15,000 years ago.[5] It is dicey to pin down the specific location further than this, but the researchers bet on China as the origin point.[6]

The dog is, beyond question, the first domesticated species. What is the backstory, exactly, that may explain how wild wolves became the steadfast dogs that offer to our species friendship and joy?

RELATING, NOT TAMING

Let's imagine the hunter-gatherers of China around 15,000 years ago. We may compare two scenarios to explain how they began to relate with wolves in ways that eventually led to the domestication of the dog. Both scenarios depend on the humans' evolved familiarity with the other animals around them, a familiarity that bred animal-watching and -tracking skills of the keenest order. Not all animals, people noticed, behave in identical ways, just as people themselves vary in their temperament and their habits. In the case of wolves, some are more aggressive and others less so; some flee or snarl at the approach of humans, whereas others are less reactive. Some wolves are drawn in close to camp by the smells of cooking meat or the hope of dragging off a bone, whereas others keep a fearful distance.

Version A. One day a member of the group, perhaps an older child, finds a wolf pup that somehow got separated from its mother. The pup is old enough to survive without its mother's milk, and when the child brings her back to camp, she grows up fairly comfortable in human surroundings. When the time comes, the young adult wolf creeps out of camp and obeys her instinct to mate with a male wolf, then births a litter. The pups are acclimated to humans from the first moment that the light strikes their just-opening eyes. In camp, people enjoy the contact with the pups, who stay near, grow up, and produce litters of their own.

Some of the hunters in camp realize that with these keen trackers by their side, the hunt goes more easily and is, on average, more successful. The whole group—men, women, and children—now encourages these not-so-wild wolves to hang around. Generation after generation, the wolves become more attuned to humans and their ways. Domestication is on its way, and the eventual result is the loyal dog that we know today.

Version B. An unusually fearless female wolf begins to prowl around a human camp, and even follows some distance behind one of its hunting parties. Wolves are pack animals, with keen social instincts and a preference for hunting by day. The group-oriented behaviors of humans thus have a familiar feel for this female.

It's hard work to get enough to eat, and the female's innovative approach pays off for her in terms of extra calories and protein—and in greater reproductive success. The pups of this well-nourished wolf enter the world a little meatier, a little more robust, than their counterparts, and with a genetic predisposition to explore. Like their mother, they approach human settlements; their proximity to humans acts as a sort of protective buffer against predator attacks, and the wolves' mortality rates decline. In a number of ways the wolves' lives improve when they attach themselves to humans. The humans benefit, from cooperation with the wolves during the hunt, and because people derive emotional pleasure from being with these animals.[7]

Humans tame the animals, or *Animals choose to become domesticated.* These pithy slogans sum up version A and version B, respectively, of how dogs went from living wild to joining up with human groups. In anthropology's study of domestication, pride of place these days goes to version B. That humans imposed their superior will on animals is less likely a scenario than one of partnership that brought benefits to all parties involved.

In *The Covenant of the Wild,* Stephen Budiansky emphasizes how natural it is to resist the second scenario in favor of the first: "The limit-less power of human invention is something we simply take for granted; it seems obvious that we 'imposed' domestication on animals, just as we imposed our will in countless other ways on the world around us."[8]

Budiansky is right. This view *is* prevalent. Sometimes it is con-veyed with a sort of tired resignation, as if we humans, and the ani-mals around us, must endure the legacy of the havoc our ancestors wreaked on creatures of the natural world. The writer Michael Korda, in reviewing a trio of books about horses, refers to the job of "a horse-man" in such resigned terms. He (or she) must "get the horse to accus-tom itself to *our* world, since that's where it lives, for better or worse, now that we have domesticated it."[9] *Our world.* But the human world hasn't been separate from the horse's world for many millennia; it developed in interaction with the horse's world, through the grand sweep of animal-human interactions.

On the topic of domestication, Korda's prose is mild compared with Paul Shepard's. Fairly spitting with anger, Shepard conflates domestication with exploitation: "The benefit to animals of being domestic is fictitious, for they are slaves, however coddled, becoming more demented and attenuated as the years pass."[10]

The science simply does not square with Shepard's view. To rec-ognize that not all horses, or all dogs, are treated well by their care-takers, or to agitate against abuse of animals in the farming industry, is a right and necessary thing. The burden of human atrocities against animals cannot reasonably be laid at the door of domestication, how-ever. When Budiansky first embraced the notion of animals choosing domestication, the response from some quarters was akin, he writes, to adding "a scientific dimension to an emotional truth." This hap-pened because "people who work with animals knew this in their bones all along." They knew that *mutuality* was the key to under-standing what happened when animals joined up with humans.

Why have scientists adopted this sea change in outlook? For one thing, Budiansky points out, it's hard to ignore the divergent historical trajectories of domesticated and nondomesticated animals: "Domestic

dogs, sheep, goats, cattle, and horses far outnumber their wild counterparts. The global populations of sheep and cattle today each exceed
one billion; their wild counterparts teeter on the brink of extinction."[11]
This observation hits home for me, because of my work with homeless cats. The astounding reproductive success of our pets has caused a
tragic imbalance, far fewer homes available than there are animals that
thrive in close contact with humans. This fact lends weight to the idea
that evolution favored those animals more likely to associate with
humans for reasons rooted in the animals' own survival and reproduction, even apart from any benefits to humans.

At some point, thanks to human management, populations of the
wild wolf *Canis lupus* changed enough to become a biologically distinct species, the domestic dog *Canis familiaris*. Yet the sharp distinction that taxonomic science requires—an animal is *either* a dog *or* a
wolf—may in the real world be somewhat fuzzy. In the 1990s, scientists tweaked the classification *Canis familiaris* to make it *Canis lupus
familiaris*, a recognition of the ways in which the dog is "a mere variety of the wolf."[12] And although most early dogs were distinct in size
or skeletal form from wolves, so much overlap in skeletal form and
even body size occurred that today's experts worry about using
anatomy alone as a guide to the domestication process.[13]

A decrease in aggression and an increase in docility may have
marked the first protodogs compared with wild wolves, and if that's
so, the true origins of the domestic dog will be masked by a search for
bones. Researchers Robert K. Wayne, Jennifer A. Leonard, and Carles
Vila note that protodogs, in following nomadic humans, may have
"utilized food from refuse and carcasses, and in turn may have provided humans with early warning systems or defended occupation
sites from other carnivores." This nomadic existence would have "led
to differences in reproductive timing or mate choice" in the dogs as
opposed to wolves, another set of changes that fossilization would fail
to reflect.[14]

Physical and behavioral changes marked the wolf-to-dog transition, but it was the emerging benefits for all involved that drove the

process itself. The fruits, or, more aptly, the flesh, of cooperative dog-human hunting probably tops a shortlist of benefits. Extra fat, protein, and calories garnered from bringing down an extra animal now and then are the stuff of a simple evolutionary calculus for any species: for dogs and humans alike, extra calories make extra energy that in turn allows for enhanced rates of reproduction. Taking a step back, we can see that cross-species stalking and ambushing of game must have selected for individuals who comprehended fairly subtle signals across species lines. Today's experimental science can tell us a lot about these communication processes and how they evolved.

STUDYING TODAY'S DOGS

Thanks to my obsession with perusing academic journals online during my lunch hour, I noticed the trend early on. About ten years ago, the scientific study of dog cognition and communication went red-hot, especially among those of my tribe, the primatologists. Everyone wanted to know, are dogs as smart as monkeys and apes? Of course, the answer depends on the experiment or task at hand. In one area, dogs outshine even clever chimpanzees. Researcher Juliane Kaminski of the Max Planck Institute in Leipzig, Germany, says flatly, "When it comes to understanding human behavior, no mammal comes even close to the dog."[15] Kaminski is one of five scientists from MPI to test dogs experimentally for their understanding of human communicative cues—and to relate the results directly to domestication.

In a series of trials, MPI experimenters hid food from dogs. The dogs were no canine cognitive superstars, but garden-variety household animals. (The MPI website invites people living in and around Leipzig to sign up their dogs, all breeds welcome, for participation in the research.[16]) In these particular experiments, young dogs, ranging in age from six to twenty-four weeks, were used.

The basic setup went like this: Experimenter A held a dog by the collar, while Experimenter B sat nearby on the floor and, out of the dog's sight, hid food under one of two identical cups. (Both cups had been put into contact with food in order to neutralize the role of odor.) Next, Experimenter B pointed to the correct, baited cup, and looked back and forth between it and the dog. Experimenter B then released the dog.

Variations on this basic setup were performed, in terms of the cups' placement and the experimenter's precise signal to the dog. The results were striking. The dogs consistently chose the baited cup. Even six-week-old puppies could read the humans' communicative signals!

In similar experiments reviewed by the researchers, neither apes nor wolves performed as well as the dogs. This superior performance was true even with wolves hand-reared by humans. That the dogs outperformed wolves as well as apes lends credence to the idea that domestication shaped the dogs' abilities to communicate with humans. Though I would have preferred that even younger dogs be tested by the MPI researchers, I take their point that the behavior of six-week-old puppies is not likely to have been shaped extensively by interaction with humans. "From an evolutionary perspective," the scientists conclude, "the results support the idea that dogs' communicative skills are a special adaptation and the result of selection processes during domestication."[17]

Experimental evidence cannot discern whether our ancestors chose wolves more attentive to human cues or, instead, wolves more attentive to human cues chose in effect to domesticate themselves (our scenario A versus scenario B). Still, the research results do cohere with anthropology's favorite explanation (scenario B). A communicative connection between dogs and humans was of benefit to both species, and came about through a process of mutual will and choice. Once it emerged, it led to new ways of humans' being with animals.

DOGS AND RELIGION

Twelve thousand years ago, at a site now called Ein Mallaha in northern Israel, people lived in stone settlements. They stored grain in pits, but also hunted gazelle, and fished. No sheep or goats were kept in the village, as far as archaeologists can tell. Yet a woman was buried at Ein Mallaha with a puppy resting under her left hand. What an evocative image! The woman's hand on top of the puppy may symbolize for us the link shared by the two individuals in life.

I cannot come up with any credible explanation for the double burial *other* than shared affection, and in this I echo the anthropologist Darcy F. Morey's view. After a review of dog burials throughout prehistory, Morey concludes that the "buried evidence is a strong and compelling indication of the relationship between people and dogs . . . when family members die, we usually bury them."[18]

Family members! Here may be the origins of the human embrace of animals into the hearth and household. Throughout prehistory, Morey tells us, dogs and people shared the grave—not routinely, but not exceptionally, either. When they did, were they meant to share in death what they had shared in life? In some cases, the burials may indeed attest to a religious sensibility. Yes, it could be that dogs were buried from affection and shared bonds alone, in a wholly secular context. But in some prehistoric burials, people, dogs, and dog effigy pots and pipes are interred together. For Morey, this strongly implies a perception of spiritual qualities in the dogs.

Dogs in some ancient societies, then, may have enjoyed venerated status, as well as the fruits of human affection and loyalty. We must not romanticize this relationship, however. Sometimes people broke the dogs' necks or killed the dogs in some other way, in order to inter them with the people who had died. Was this sacrificial status an indicator that dogs were entwined in sacred rites? Possibly yes, if a parallel existed with human sacrifices that were intended to appease the gods—and that brought honor to the victims. "Regardless of how they died," Morey concludes, "[dogs] were buried with the kind of care that signifies friendship, symbolic of a projected afterlife in the 'spirit world,' more regularly than other animals."[19]

What *about* these other animals?

CATS, AND MORE

Ancient Egypt is the most glorious and famed cat-oriented culture of them all. By about 4,000 years ago, the cat had been domesticated in Egypt. Over time, cats became suffused with sacred meaning in Egyptian culture. First, cats became associated with gods; then, at the height of sacred-cat culture, cat statues were made into temple offerings—and the cats themselves were mummified.

But Egypt is not the *first* evidence of human-cat relationship. Over 5,000 years earlier, closer to the time period of early dog domestication, a young cat and a person were buried next to each other in Cyprus at a place called Shillourokambos. The person was buried with polished stone, axes, ocher, and flint tools near a pit containing two dozen marine shells. Only a little over a foot (40 cm) away, an eight-month-old cat was interred. No docile animal, this particular feline was a wildcat.

Like the burial with the woman and the puppy in Israel, this close association between a person and a feline speaks volumes. Without a visible "cat culture" like that of the Egyptians, and for that matter before domestication produced a creature visibly different from a wildcat, a cat-human bond transcended death. I like to imagine the Cypriot person walking the streets of his or her village, looking forward to seeing the young cat and perhaps interacting with it at home. It's impossible to map any specific emotional tenor onto that relationship, and indeed, it's a little eerie to think that—just as with ancient dogs—the cat may have been killed prematurely in order for the double burial to go forward. (A sacrificial scenario is clearly supported in the case of Egyptian cat mummies, where scientific analysis reveals the deliberate killing of the cats.)

To us today, to devalue so completely an animal's right to live may seem incompatible with a felt bond between humans and cats. But in the remains from early Cyprus, we have a glimmer, if only a faint one, of a sacred connection among people's lives and animals' lives—and deaths. The remains testify that a human-cat relationship "during the eighth millennium in Cyprus was not restricted to the material benefit of humans but also involved spiritual links."[20]

The "mutual choice domestication" that began in an era before history still impacts the day-to-day lives of animal lovers. We embrace our pets in our daily routines, our vacations, and our religious lives. Our physical and mental health may increase when we care for, and engage with, our cats and dogs. That our animal companions also thrive on this shared intimacy is an evolutionary gift that stems directly from our ancestors' pairing up with wolves, or with wildcats. When we observe in a museum the vital artifacts of ancient cultures, perhaps the hunting implements of Paleo-Indian groups or symbols of the religious burial rites of the ancient Egyptians, we sense the link between societies of the past and those in which we live today, societies almost universally technology-oriented and religion-saturated. But for those of us with animal-filled homes and backyards, we need not step out of our own domestic space to feel the past at work in the present.

Other species, too, joined up with ancient humans. Most of us today tend to be less sentimental about these others. Outside of children's tales, comparatively few people muse about the loyalty of the goat or the cleverness of the sheep, or wax romantic about shared emotional states with a cow or pig. (I don't mean to exclude these species from the sphere of human affection. The world over, people *do* bond with these animals. As I wrote this section, I stumbled across a video clip on the Internet, of a ram named Nick who lived happily indoors with his human family in Wales. Together, humans and sheep ate, traveled in the car, and watched TV. Mutual affection was more than evident.)[21]

Yet the impact of these animals on our history was tremendous. When sheep, goats, cattle, and pigs joined the circle of domesticated beings, the pattern of human productivity, labor, and diet changed forever. These animals carried burdens, or gave meat or milk or wool, and entered into systems of barter and sale. Were they strictly utilitarian creatures, or can anthropology reveal ways for us to think about the animals' emotional or religious resonance to people early on?

In Turkey, at the village of Catalhöyük around 8,000–9,000 years ago, a man was buried with a lamb. The man's skull had been crushed. He was buried on his right side with his legs flexed to the chest. Those who buried him placed atop his chest a worked bird bone and a flint object, and, behind his shoulder, a bone point. Archaeologists Nerissa Russell and B. S. During describe the lamb's peculiar position: "It lay in a contorted posture on its left side: its hind legs extended straight behind it at roughly a forty-five degree angle upward, while its front legs were close together and extended straight up vertically, twisting the forequarters so that the legs were essentially at right angles to the body."[22]

This odd position, together with the fact that the man is separated from the lamb by a type of mat or blanket, suggests that human and animal were purposefully kept separate in death. Other clues, though, point in a somewhat different direction.

Catalhöyük was, at this time, a village teeming with domesticated sheep and goats. The animals were raised for meat. To bury them, much less in human graves, was an uncommon act. Yet this lamb was interred, together with the man, in the traditional burial place: a pit dug beneath a house floor. Other people were subsequently buried in the same precise location. This single clue, the use of a domestic burial ground, hints strongly that the lamb meant more to someone than a routine flock animal would have meant. Russell points out that the etymological roots of "domestication" imply "bringing into the household." And indeed, she writes, the lamb "was placed with the ancestors and perhaps became an ancestor itself. This seems a strong statement of kinship with the man buried beside it, the three people later buried above this man, and the people who continued to live in the house above." Yet it's not as if the lamb was a fully integrated member of the family, either: "[T]he ambivalence of its position, held awkwardly apart from the man it was buried with and carefully avoided by the subsequent burials, suggests that this kinship was not uncontested."[23]

What did the lamb mean to the man with whom it was buried? Or to the man's family? Russell and During consider a number of possibilities. Perhaps this person enjoyed a kind of intimate emotional relationship with the young sheep in the way that shepherds often do, especially when sheep are separated at young ages from their mothers.

Might there have been religious overtones to the double burial? Russell and During don't believe that the lamb was meant as food to fuel the man's afterlife; a whole animal is not needed to accomplish that. Nor is it likely that the lamb was intended as some kind of spirit guide in the afterworld. That idea is not without a certain appealing logic, but evidence for it is sorely lacking. Sheep at Catalhöyük are not routinely buried with people, and there's certainly no cult of the sheep going on!

Two conclusions seem clear enough. This lamb meant a great deal to the person buried with it. If we go any further than that, we enter the realm of reasoned speculation. Hypotheses that posit religious sensibility in human-animal relating simply cannot be disentangled by means of a "subject pool" of a single buried lamb.

Notably, the lamb burial occurs at Ground Zero for domestication. More than any other region, the Near East's Fertile Crescent, which encompasses parts of Turkey and also Syria, Iran, and Iraq, has yielded evidence for the domestication of animal and plant material. Melinda Zeder and Brian Hesse note the "remarkable array of today's primary agricultural crops and livestock animals" that originated in the Fertile Crescent: wheat, barley, rye, lentils, sheep, goats, and pigs.[24]

The study of goat-and-sheep domestication, a surprisingly vital area of research, comes replete with its own vocabulary. The progenitor of today's domestic sheep was the Asiatic mouflon, which lived in the mountains of the Fertile Crescent. Its counterpart in the goat world, with a similar geographic distribution, was the bezoar. Zeder and Hesse detail a step-by-step transition from human hunting of bezoars to full-blown captive breeding and management of goats.

The timeline begins at about 40,000 years ago. Both Neanderthals and early modern people lived, hunted, and gathered in the Old World at this time, though not necessarily in the same specific areas at the same time. We tend to envision robust Neanderthals grappling with massive cave bears, but in the Zagros Mountains of Iran and Iraq, their hunting traditions focused heavily on mouflon and bezoar.

When *Homo sapiens,* with their superior planning and language skills, began to overtake the great Neanderthal hunters, they zeroed in on the very same species. Our ancestors, then, must have attended closely to the habits and behaviors of the wild goats and sheep they ate, for, as we have seen, a successful hunt is based on close observational and tracking skills.[25]

By around 10,000 years ago, people began to manage goat herds, right in the region where the wild goats had always lived, and were hunted. Population demography, via the analysis of bone remains, led Zeder and Hesse to recognize "a clear pattern of selective kill-off of subadult males and delayed slaughter of females that matches culling patterns in modern domesticated herds managed for meat." In other words, subadult males make good eating, whereas females are kept alive longer, in order to boost the herds' reproductive rates. The skewed age-sex profile in past goat herds can only be explained by clever decision-making on the part of our ancestors.

Lost in time are the precise steps that lay between bezoar hunting and goat-herd management. We can reasonably imagine them, though. Initial hands-on control of wild goats may have been relatively light, amounting perhaps to a kind of managed hunting strategy. This stage would have morphed into a new one where control of captive goats was performed *outside* the natural range for the species. Once full domestication was under way, the goats' anatomy began to change, too: horn shape, limb length, and coat color all altered. (Zeder and Hesse back off from a long-accepted marker of domestication: overall reduction in the animals' body size. Body size, they say, responds flexibly to many aspects of the environment outside human influence.)

We might wonder, why would the goats agree to this degree of human control? What was in it for them? Our scenario B—the one based in mutual choice and benefit for animals and humans alike—requires a long-term, multigenerational process that featured some animals' first steps at experimenting with a life near humans. Driven by the animals' reproductive success, the process would then have fed on itself. In this regard, Budiansky's remarks about sheep or cattle probably apply to goats as well: "These animals would have gained measurable advantages from flocking with humans—being able to scavenge campsites or grainfields and live under a shield that guarded them from other predators."[26] No factors are more vital to reproductive success, or more likely to cause an evolution in animal behavior over time, than a heightened ability to attain food or to avoid becoming it.

A trajectory similar to the one outlined by Zeder and Hesse for goats probably occurred, too, with sheep. Certainly by the time of Catalhöyük, as we have seen, sheep lived in human-managed herds. For goats and sheep, the context differs from the one in place when the dog was domesticated. When wolves chose to frequent human-populated areas, it was not around ancient villages, but at camps of nomadic hunter-gatherers. The wolf-to-dog transition significantly predated houses or crops. With other animals, it is impossible to generalize across the board about the nature of the societies in which humans lived. Yes, the origins of domesticated plants and animals (excluding the dog) date to Near Eastern villages before 10,000 years ago—but beyond this vague kind of summary, things get complicated. The progression from hunting-and-gathering first, to plant domestication and agriculture next, and then on to settled communities, with people at last erecting permanent buildings, is a sequence at once tantalizingly tidy *and* roundly discarded by current anthropology.

For one thing, as the anthropologist Peter J. Wilson notes, a built living environment precedes agriculture by between 500 and 2,500 years depending on the location. Along the Euphrates River (in today's Syria), people lived in semi-subterranean buildings for two millennia before they began to cultivate crops! At Jericho, an ancient town well known to anyone versed in the Bible (it's where "the walls came tumbling down") hunters and gatherers were sedentary. "In the Near East," writes Wilson, "and in fact everywhere else where domestication was indigenous rather than imported, an architecture preceded cultivation."[27]

That people emerged from permanent houses only to set off to hunt game or gather fruits and nuts is not a fact celebrated in popular science. It puts the lie, though, to an equation of prehistoric hunting and gathering with constantly roaming bands. Even 20,000 years ago, Ice Age hunters erected a semipermanent structure of mammoth bones. Wilson, though, means to focus on changes of a more sweeping nature, those wrought by the experience of creating and living within permanent structures. When people began to settle, Wilson believes, they also began to think in new ways. Not only did they measure and construct their buildings with new rigor, but their very patterns of reasoning were affected because units were now organized into composites: rooms in a house, and houses in a settlement, with doors and gates allowing movement from one place to the next along well-trodden walkways. Pedestrian paths become routinized; people were more easily counted and controlled. As we have seen with the lamb burial at Catalhöyük, people begin burying their dead beneath their domestic living spaces. And as Wilson remarks, "People domesticated themselves first. They made themselves at home."[28] From this simple base, they went on to invent pens to hold animals captive, and patches to plant crops.

It's a good argument. Whether it holds in 100 percent of cases, I don't know. After all, the dog became domesticated well before anyone was living in a built environment. And in Israel, people's diet already included wild wheat and barley by around 23,000 years ago. The BBC reported this discovery in 2004 as "farming origins gain 10,000 years," but this is an overstatement, as agriculture is more than the manipulation of wild grasses. The BBC journalists went on to say that humans made "their first tentative steps towards farming" at this time. While perhaps a little hyperbolic, this statement is closer to the truth; the Israel plant data do blur the line between strict hunting-and-gathering and a type of foraging that might have led to experimental agriculture.

It's productive to consider an architectural revolution instead of only an agricultural one. Did the built environment of the Fertile Crescent villages attract sheep and goats closer to humans? It's hard to know, but settled life does lead to greater population density, to more men, women, and children going about their daily behavior in one place. And maybe the stable food sources associated with such a place were simply irresistible to animals—especially to young ones.

Budiansky explores what he calls the "juvenilization" of domesticated animals, the "tendency to treat humans as members of their own species, as opposed to either predators or prey."[29] As anyone with a toddler or teenager can attest, in our species, too, it's the young who are inclined to be experimental—if not outright reckless. It seems to be a mammalian trait, across the board. If village life, with its attendant food sources, few competitors, and fewer predators, attracted young animals *and* led to their relatively rapid reproduction, this state of affairs could easily have resulted in "a powerful selective force in favor of animals that reach sexual maturity earlier."[30] Did an intense feedback loop govern animals' willingness to be domesticated: an animal's relative fearlessness and tendency to explore human settlements; the comparative ease and safety of life near humans leading to enhanced reproductive success; and the passing of these behavioral and reproductive tendencies along to the next generation, so that animal populations became even further juvenilized and human-oriented?

Whatever the exact steps that made up the domestication process—a process variable according to the species of animal and the lifeways of the humans in question—the key fact is that *both animals and people* took them. This doesn't imply some sort of romantic negotiation process. Just as I feel nothing but scorn for the notion that hunters and the hunted may become locked in some sort of symbolic embrace, causing the hunted to give up its life in a daze of regretful-but-courageous sacrifice, I reject any implication that prime-age male goats of the ancient managed herds cooperated in their own culling and thus hastened their own deaths. Nor do I think that Egyptian cats presented themselves to humans in some kind of weird death pact, in advance of mummification. The science behind animal domestica-tion, rather than suggesting some kind of mystical animal mar-tyrdom, guides us to think of the reproductive and evolutionary advantages of human-animal cooperation.

This evolutionary argument gains strength from one last line of evidence: not all animal species wanted in.

THE MOOSE RACES?

Conjure for a moment an alternative universe. In this unreal place, raccoons are given pride of place in people's homes, with soft beds to rest on and amusing toys to play with. They are walked on leashes in the fresh air, and taken to the vet for medical checkups. Horse-racing is nonexistent in this parallel world, but that doesn't mean the Kentucky Derby is dead—it's just run with moose. Preteen girls dream of the chance to ride free on a moose. They collect stuffed moose and books about courageous girls who thrive on moose adven-tures. Meanwhile, the girls' parents, out for a gourmet meal, indulge their palate with gazelle cheese, a delicious delicacy favored by fine restaurants in all the world's capitals.

Raccoons, moose, and gazelles are just some of the animals that people *tried and failed* to domesticate. "The extraordinary high failure rate of man the domesticator" is as much a part of our history as is taming fire.[31] *Why* some animals opted out would make for a fasci-nating study in itself. For now, it's enough to add human ineptitude to our cache of clues that point to animals' cooperating with humans in their domestication. It's a humbling piece of our history, in its way, but also a beautiful one: some animals chose to accompany us on our evolutionary journey as hunter-gatherers, farmers, village-builders, and, later, city-dwellers. It may be fanciful to wonder this, but wonder I do: Is it a choice they regret?

From the smorgasbord of animal domesticates, there's one final species I want to consider. This animal, perhaps more than any other, has paired not only with our human society but also with our spirit. An uncommonly graceful animal, the horse has given to humans the power of speed. When a person and a horse move as one over the landscape, always with purpose and sometimes with joy, the result is elegance.

In *Horse: How the Horse Has Shaped Civilization*, J. Edward Chamberlin reminds us that the cave walls of Chauvet and Lascaux shimmer with horse images:

> This is the early history of humanity, and horses are consistently present. The tradition includes carvings in ivory all across Europe and Asia; decorative pendants representing horses' heads, often made from the bone in a horse's tongue; and tools such as shaft-straighteners and spear-throwers with horse images on them. Along with the cave paintings, these celebrate the spectacular importance of horses—running free in the fields of the imagination—to people who defied the soulless utilitarians of ancient times.[32]

Early *Homo sapiens'* preoccupation with the horse included hunting it for food; adorning cave walls with its image; and controlling it to develop new heights of hunting and transport. In the Fertile Crescent, horses were domesticated well after the time period for goats and sheep. In Iran by about 5,000 years ago, horses are associated with human sites (and not just as the remains of hunting forays). But the most spectacular early finds for the horse come from Central Asia, in the country of Kazakhstan where the Botai people lived.

Impressive evidence for horse domestication comes from this region, the very heart of the wild-horse range:

- At a site called Krasnyi Yar, a circular horse yard was unearthed and dated to 5,600 years ago. It appears to be a type of corral.
- At the site of Botai itself, before 5,000 years ago, people were manufacturing rawhide thongs. Sandra Olsen, an expert on horse domestication, notes that the thongs "might well have been used in a wide range of activities relating to both riding and capturing horses."[33]
- At Botai, two men, a woman, and a child were buried with fourteen horses surrounding them in an arc. Once again we are confronted with the expression of human-animal emotional connection in the form of ritual killing of the animals.
- Horses and dogs were sometimes buried together by the Botai people. Olsen remarks that "the ritual association of dogs with horses evokes the close connection between these species among steppe peoples today, where dogs are used alongside horses to hunt game and also to help herd domestic horses."[34]

I can think of no better descriptor for the Botai culture than *intensely horse-oriented*. Horses provided the Botai with food, drink, clothing, and material for tools—but like people everywhere, the Botai were about more than survival. The presence of ritual sacrifice once again hints at cross-species emotional connection, and perhaps religiosity. Human emotion seems to compel us, across time and space, to cement our bonds with animals by joining them in death as much as in life.

As Budiansky points out, it's human nature, too, to succumb to sensational claims of humans' power to tame the wild horse. For any American raised on tales (and movies) of pioneers and lawmen "taming the Wild West" via the horse, it's a familiar theme. In this light, it's worth remembering that our mutual-benefit scenario of animal domestication doesn't preclude all aspects of taming. In the case of the horse, evidence points clearly to an extreme degree of human control. By about 3,500 years ago, horses were harnessed to the wheel and the cart. Chamberlin comments that, unlike the oxen, asses, and dogs previously used to transport burdens, the horse offered speed and style.

Here, then, was a magnificent animal of a nature arguably different from all the others. It ushered in long-distance migration for hunting and warring, and for trading across cultures, in ways previously unimaginable. Chamberlin's book reviews the Great Horse Cultures, and details the myriad ways that the human-horse relationship changed them. Budiansky, as usual, comes at the issue from the opposite direction, noting that domestication probably saved the horse from extinction. A combination of environmental change and human hunting had caused several declines in wild-horse populations throughout Asia, North America, and Asia. The emerging association with humans reversed that trend.

THE WILD AND THE TAME

Are there any truly wild horses left in the world? This question is as much one of definition as it is of horse biology. And it speaks to this chapter's subtext, the fatal flaws in either-or thinking about animal domestication. Using data from anthropology, we have exploded a number of supposed either-ors: hunting-and-gathering *or* farming (people sometimes hunted, gathered, and experimented with grasses, all at once); nomadic hunting *or* settled farming (people sometimes lived in sedentary villages and yet still hunted); presence *or* absence of domesticated animals (people sometimes managed wild flocks of animals, but still buried their dead along with a single animal); the utilitarian *or* sacred aspect of being with animals (people embraced some animals in sacred ceremonies even while using others for labor). Let's contest one final dichotomy: wild *or* domesticated animal.

It's easy enough to assume that the domestication of dogs, cats, goats, sheep, horses, and all the rest not only caused revolutionary breakthroughs in human society, but also created essentially new animals. In some ways, domesticated animals really *are* different. Compared to their wild counterparts, they are more able to succeed (as populations) in evolutionary terms, and more prone (as individuals) to forge emotional bonds with humans. My own experience, however, supports what the scholars of animal domestication say: it's not that simple.

The monkeys and apes that I study, along with innumerable other species of mammals and birds, year by year are forced into closer contact with human populations. The buzz of chain saws fills the rain forests' air as logging companies cut down acres upon acres of trees— the homes of monkeys, apes, birds, and many other animals—for lumber. The logging roads open up the forests even more to human encroachment, bringing with it disease and poaching. Sanctuary areas, from Gabon's national parks to America's, succumb to the pressure of more and more nature-seekers; the animals become victims of the tourist vans that press in on them, the cameras that flash at them, and the foods that people hurl at them. Day by day, the wild alters, and the animals with it, because of human population pressure.

But what *is* the wild? Does the wild exist only in Wyoming's vast Yellowstone National Park or Kenya's huge Amboseli? Or can it be found also in Manhattan's smaller Central Park? Do our own tiny backyards carry a trace of the wild? Are the wild deer and bear that take occasional strolls down Main Street, overturn our garbage cans in search of food, and lumber up onto our backyard decks somehow not so wild after all? Or are these animals a hybrid of pure-wild and semi-domesticated, because they are accustomed to eating humans' food and tolerating humans' presence? Isn't it odd that we rarely think to include creatures like squirrels or trout or wasps and mosquitoes in our tales of encountering the wild? Are they not as wild as bear and bison?[35]

The prehistory of animal domestication reveals to us the potential for sharing a human world that lurks in many wild creatures. And it works the other way, too. Dogs may still mate with wolves, after all. And cats may morph from sweet purriness to lashing fierceness in a heartbeat. For anyone who has not witnessed this terrifying feline transition, let me attest to how fast and furious it may be. A beautifully patterned gray cat, badly injured, was abandoned at the feral colony we care for one evening in 2008. Cat carrier and water bowl were left there, as if to say, *I cannot care for him, but I know you will.* We named him Patrick, and we soon found out that he needed vet care, antibiotics, and long-term treatment with healing creams for wounds on his head and legs.

To our relief, Patrick allowed us to swab his open wounds, even when it must have caused him discomfort (or perhaps real pain). One day, fairly far along in his recovery, I succumbed to an urge to gift him with fresh air. Outdoors my husband and I went, gently pulling Patrick on a newly bought leash. To our shock, he immediately slipped out of the collar and shot across our yard, across the road, and into the woods. Fortunately for us, alluring smells (we speculate) stopped him in his tracks. My husband tackled him, I ran for a towel to envelop Patrick, to calm him and hold him still, and we dispatched our daughter to retrieve a cat carrier.

The towel-calming strategy failed. Swaddled in cloth, pressed hard into the ground, Patrick turned into what is termed a "wild animal." It was an awe-inspiring sight. Eyes full of fear, he screamed and tore at us with teeth and claws. The two of us, strong adults, were mightily challenged to hold down a cat that weighed less than twelve pounds. This "tame" cat nearly overpowered us. (Nevertheless, we did squeeze him into the carrier, won back his trust, and completed his medical treatments.)

As Patrick teaches us, the supposed wild/domesticated dichotomy is better thought of as a continuum. The behavior of individual animals is often, depending on the species, fluid, flexible, and responsive to experience, rather than hard-wired, and it is that tendency which helped to write human history.

To understand that animal-human relating, with all its gloriously varied emotional and religious overtones, goes back so many millennia is a curiously empowering knowledge. It is knowledge that brings with it serious responsibility. Animals joined with us when it helped them to do so; the onus is on us, evolved creatures that we are, to work harder to see to it that our cross-species association remains (or, in some cases, becomes) a positive.

The willingness of wild creatures to join with humans has helped to propel *Homo sapiens* in new directions: beyond the at-a-distance admiring of animals, painting of animals, and hunting and spirit-worshipping of animals, toward a new intimacy born of living with animals side by side and day by day. This daily communion in turn ushered in new opportunities for humans to represent their worlds— their sacred worlds as much as any others—through animal symbols.

4

Cat Mummies and Lion Symbols

• • • •

A LION CHRONOLOGY

Prehistoric lions: On the walls of France's Chauvet Cave

Very old lions: At the Sphinx in ancient Egypt and the Lion
Gate of Mycenae in ancient Greece

Guarding lions: At Beijing's Forbidden City, Nelson's Column
in London's Trafalgar Square, and the New York Public
Library

Today's lions: The Detroit Lions, *The Lion King,* "The Lion
Sleeps Tonight," *A Lion Called Christian*

ANIMAL SYMBOLS SATURATE American popular culture.
Some have been around for centuries. Churches at Christmas
depict the Nativity scene complete with live lambs or plastic ones.
Wall Street is bullish, bringing a smile to investors' faces, or bearish,
eliciting a frown. For well over a century, presidential election years
feature Democratic donkeys and Republican elephants.

Christmas plays track Bible verses, pretty much straight on. But
why should a bull be positive and bear not so? All right, bulls charge
forth (like a bull in a china shop!), while bears lumber about and are

more cautious by temperament. But is a bear not as strong as a bull, and smarter to boot? Couldn't we argue the case either way, of which animal is more powerful? That's the fascinating thing about animal symbols, how they make meaning by tradition and nuance as much as by direct connotation. Even the donkeyfied Democrats express an in-the-eye-of-the-beholder grasp of political symbology: "The Democrats think of the elephant as bungling, stupid, pompous and conservative—but the Republicans think it is dignified, strong and intelligent. On the other hand, the Republicans regard the donkey as stubborn, silly and ridiculous—but the Democrats claim it is humble, homely, smart, courageous and loveable."[1]

The media explosion of recent decades has pushed animals to new popularity. Television tigers urge us to eat cornflakes frosted with sugar, while a big-oil corporation once swore it could put a tiger in your automobile's gas tank. Even in TV land, though, it's not all about brute power. Purring cats meow for their preferred dinner products. Ducks hawk insurance with honking acronyms, and geckos get into that game, too.

The insurance company Geico's gecko is, in fact, instructive in its genesis. No brainchild of animal symbologists, it was created by admen when a Screen Actors Guild strike prevented live actors from doing the on-air selling. The small animated lizard touched a chord with the public, and took on a life of his own. Apparently a talking lizard ups sales! After the original incarnation, ad teams tweaked his voice and accent, switching the voice actor from American to British and morphing toward a more working-class aura. Geico online now sells gecko gifts: everything from lizard-bedecked golf balls and baby overalls to iPod covers.

Surely in the arena of sports it's a straightforward matter, with animals chosen for their positive qualities. The idea seems to be that group identity, and thus strength, of American sports teams and their fans will be boosted via linkage with the muscles or moxie of the animals chosen to represent them. Think of football's Bears in Chicago and Lions in Detroit, basketball's Mavericks in Dallas and Hornets in New Orleans, baseball's Tigers in Detroit and the Mud Hens in Toledo.

The Mud Hens?

Minor-league baseball is not a sport I follow, but the name has lodged in my mind ever since the 1970s when I was addicted to the television show *M*A*S*H*. Memorable even among the tragedies and comedies of that mobile medical unit's Korean War experiences, character Maxwell Klinger's love of the Toledo Mud Hens was a comforting constant. Yet the team and its name are no fictional device.

Since 1896, when it first played in a swampy area in town, the Toledo team paid homage with its name to the funny-looking marsh birds called the mud hens. Enthusiasm for the name, and its ability to capture the public's attention, has grown by the decade. These days, team mascots Muddy and Muddonna hire out for kids' birthday parties or company picnics, and the Mud Hens of course have a Web presence. Their official site includes links where people can hear mudhen vocalizations, which sounded like generic goose honks to my unornithological ear, and read more about the lowercase mud hens themselves.

In search of a clue to the obvious question—why name a baseball team after a long-legged, short-winged swamp bird?—I did read on (at least as far as the linked Wikipedia page). Also known as coots, mud hens are "weak fliers" and "require a great deal of effort to become airborne," but they do have "considerable stamina" once up in the sky. What better qualities, really, to lend energy and support to the image of a minor-league team with heart? (I wrote the above just twenty-four hours after two Mud Hen players were named to the U.S. Olympic team for the summer 2008 games in Beijing; even swamp birds have talent!)

Such animal symbols are everywhere. Think rock music, as another category: the Eagles, Def Leppard, the Byrds, the Turtles, White Snake, and let's not forget Atomic Rooster—or the band the

Animals. My own adolescent bedroom was sheathed in posters of the Monkees: who knows what subliminal connection was forged in my brain that pushed me toward a life of studying monkeys and apes? And now I live near Richmond, Virginia, the home of the heavy-metal band that calls itself Lamb of God. This name brings us straight back to lambs at Christmas—and to the lamb buried at the ancient Turkish village of Catalhöyük.

The domestication of animals, which led to that person-lamb double burial so long ago, was the first step in a new evolutionary path of being with animals. Now meaning-making with animals began to occur in larger networks and eventually whole societies.

THE TURKISH BULLS

In revisiting Catalhöyük, we explore the very cradle of village life. Settled around 9,000 years ago, the site was inhabited for over a millennium. People lived densely, with houses packed together and as many as 8,000 in residence at one time. Catalhöyük people cultivated cereals, but ate wild foods, too; they domesticated sheep and goats, but still wild were cattle and pigs.

Using domesticated and wild creatures alike, Catalhöyük people made meaning through animals. This fact comes clear by looking at their art, the wall paintings and statues especially. These works are studded with humans in contact or connection with animals, including bulls, vultures, and leopards. One arresting statue shows a hugely fat woman seated on a chair, perhaps a throne. A tiny head is visible between her legs; she has just given birth. Many archaeologists ascribe goddess status to this woman. Whoever she may be, she's literally enveloped in the animal world. As the archaeologist Mary Voigt points out, the arms of the chair are formed by two large standing cats (leopards, according to some scientists). The woman's hands rest on the cats' heads, and the cats' tails curve around her shoulders. The woman, whom Voigt does consider to be a goddess of some sort, is depicted in a position of control over the cats. Other statues at Catalhöyük show humans, both female and male, in dominion over other creatures.[2]

To learn about these works of art in an early village gives us a sense of evolutionary continuity, for painting and sculpture were the favorite media, too, of Ice Age cave dwellers. But Catalhöyük is just about as many millennia distant in time from Lascaux's cave art as our society today is from Catalhöyük. Evolution means change, and by the time of Catalhöyük, things were different.

The houses give the first clue. No longer seminomadic, the Catalhöyük people crowded themselves into a dense settlement, where "there are few streets and . . . access to houses and animal pens was over roofs of houses."[3] Early archaeologists believed they had found shrines, places dedicated to the carrying out of religious rituals. Newer work by the archaeologists Ian Hodder and Craig Cessford shows convincingly that, in fact, domestic activities went on routinely in the so-called "sacred" spaces. People prepared and ate food there, and made tools there. In fact, all buildings at Catalhöyük acted as domestic houses, and all contain evidence of art and ritual.[4]

In a beautiful sort of irony, what makes Catalhöyük most helpful in understanding our past is how much it reveals about the ways in which villagers grappled with *their* past. Activities, ranging from house-building to burial of the dead, were carried out in the same basic ways, and in the same locations, over time. Pattern repetition of this nature, Hodder and Cessford argue, helped the villagers to build up social conventions, ways of doing things that were carried forth from one generation to another. In the absence of systems of writing, this repetition might well have aided in commemorating the past.

Animals were central to this endeavor. People embedded cattle horns in walls in patterned ways (for example, on structures' west walls). Goat horns in one place, and boar jaws in another, were made to cluster in groups of thirteen. Hodder and Cessford think "that the heads and other cranial elements of such animals when placed in houses created memories of significant events in the life cycles of the houses or the people in them."[5]

Perhaps it was initiation rituals or funerals that Catalhöyük people carried out by using animal parts like these. Some archaeologists believe the cattle horns relate to a cultural preoccupation with "cosmic or supernatural forces," because they were painted or "intricately decorated" in the manner of Easter eggs and Christmas trees and associated with "shrine areas around them."[6] This may be so, even if an idea of dedicated shrine buildings doesn't fit with the latest archaeological evidence.

Still, we shouldn't get too lost in a world of sacred forces, symbols, or even memories. Penned atop the roof in life, buried below the house in death, some Catalhöyük animals lived and died cheek by jowl with people. This village illustrates *being with animals* in a way that mixes the day-to-day with the symbolic and the sacred. Or, to put it more correctly, the day-to-day and the sacred may well have been seamless, as *was* and *is* true for many cultures. Sacred ceremony, everywhere in the world and always throughout time, has been about emotion, a fact that archaeologists grasp as readily as theologians do. Yet no behavioral schism divvied up life into "religious" and "secular" realms: it was all just life, life in a world of ritual encounters with the sacred.

Early domesticated animals at Catalhöyük were not pets. Yet by all accounts they were bound up with the resonant ritual emotionality of the humans they lived with. And what of the wild-animal images and carefully placed horns? Those talismans of relating *beyond* domestication also speak to a new intimacy, one different in nature from the hunting, game-following, and image-painting of our earlier nomadic ancestors. Wild animals, and the rituals created around them, now inhabited people's living spaces.

It's not just at Catalhöyük. Through the larger region around Catalhöyük during the time period between 8,000 and 7,000 years ago, art and archaeological remnants of ritual attest to a veritable obsession with animals.[7]

- At 'Ain Ghazal in Jordan, included with human burials were bones of pigs, probably wild pigs; figurines of clay represent cattle, for the most part, though some sheep or goats and horselike creatures also occur.
- At the site of Kfar HaHoresh in Israel, human and animal bones were arranged to create an animal mosaic inside a shallow pit filled with ash. The species depicted in the mosaic may be a boar, an aurochs, or a lion. At this site, too, are a human skull and a headless gazelle carcass, buried together beneath the floor of a rectangular structure.
- At Cayonu in Turkey, a structure called the Skull Building contained bones of animals and humans mixed together—and blood. Most striking is a flint knife with traces of aurochs and human blood.
- Also in Turkey is the spectacular and mysterious placed called Göbekli Tepe. In the words of its excavator, Klaus Schmidt, Göbekli Tepe is a "supernova!" Animal images blanket a truly monumental temple structure. Enormous T-shaped pillars are decorated with carved lions, boars, foxes, birds, scorpions, and snakes. The date is stunning: 11,5000 years ago, many millennia before the other sites. Schmidt speculates that Göbekli Tepe was a place of ritual and "a site of pilgrimage" for people in a hundred-mile radius.[8]

At Catalhöyük, no dedicated shrines existed. Of primary importance was the house, where people performed ritual practices. At Göbekli Tepe (and some other sites), sacred space, by contrast, was now held apart from domestic space.

But the temples keep their sacred secrets. Did people gather there to worship the ancestors? To carry out funerary rites? To make ritual offerings (sometimes bloody ones, judging from the site of Cayonu) to the gods? Or were they engaged in ceremonies that the modern mind cannot imagine, because we have no parallel? If only, by some kind of transmillennial magic, we could infuse the artifacts dug up from the ground, or the architecture uncovered, with the emotions they must have engendered in the ancient world!

In reviewing the archaeological data on the sites noted here, Marc Verhoeven remarks that people came together communally around "emotionally arousing" animal symbols: "It's as if people were bringing the undomesticated and wild into domestic (and ritual) contexts."[9] I admit to experiencing an adrenaline rush when results of another scholarly work accord so well with the thesis of this book. Verhoeven points to *communality* and *animal-human linkages* as two key aspects of understanding the early art and ritual of Anatolia, but they work just as well as category markers for the kinds of broad evolutionary changes I'm describing in these chapters. From its very beginning, the expression of human religious awe in ritual was communal and laced with animal symbolism.

Verhoeven suggests there may be meaning in a set of links among *human, wild,* and *male,* so that animal-human relating was "an expression of the wild, dangerous, aggressive dimensions of the domain of nature." He thus evokes a domain where men were more active, compared to the domestic sphere of women.[10] Here, I'm wary. I don't buy into strict gender divisions in prehistory; over and over again, assumptions about the limited domestic role of women in prehistory have been challenged and overturned. Also, people's focus in the ancient world went beyond wildness to embrace domesticated animals as well.

The Near East region, known as the site of first farming and early animal domestication, deserves another claim to fame. It was here that humans cemented a relationship with animals that was at once daily and divine, at once about the domestic and about the wild. It's a story of wildness and control, of shared lives and shared deaths. The glimpse we get of this ancient world is of people struggling to *think with animals,* that phrase of Lévi-Strauss's that never loses its luster.

Death is likely to have been a prominent theme in *thinking with animals*. Animals give life to humans through milk, meat, and wool, enhance human life through labor and companionship, ward off the unknowability and danger of the wild by bringing it closer or under human influences. Yet animals die, of course, just like we do. They take on sacred resonance when they pair with nonmortal beings (gods) that defy death, or when they pave the way for an afterlife.

At Göbekli Tepe is a sculpture of a half-human, half-lion figure. Klaus Schmidt calls it a sphinx, and marvels that it precedes the sphinxes of ancient Egypt by many thousands of years. Yet this strange figure is also resonant with what came before. Half-human, half-animal creatures were no strangers to our Ice Age ancestors: think back to the bison woman from Chauvet, or the bird man from Lascaux.

When Schmidt evokes an analogy to Egypt, he invites us to consider a new height of fused animal-human images—and a new level of preoccupation with the nexus of humans, animals, and death.

Let's start with a cat mummy from Egypt. A small figure lies on its side, with paws slightly curled. It is tightly bound in cloth, and in one way looks stiff and unnatural. Yet the position of the paws conjures, too, an image of the cat at play with a human companion, or running with quick grace through a sun-baked courtyard.

Why was this cat preserved as a mummy?

In the words of Daphna Ben-Tor, curator at the Israel Museum in Jerusalem, animals were associated with "every aspect of the ancient Egyptian world," and the Egyptians' "special relationship to animals was essentially religious."[11] It's not just that animals symbolically represented gods or goddesses, but that animals, the flesh and blood animals people came to know in their daily rounds, were thought to represent the sacred qualities inherent in gods and goddesses.

Bastet, the Egyptian cat goddess, was portrayed as a cat-headed woman. (Type "Bastet" into an Internet search engine!) She was worshipped especially in and around the town of Bubastis. In that region, temples teemed with cat sculptures.[12] At times fiercely war-oriented, at others times linked with fertility, Bastet's nature was complex and multifaceted, as is apt for one associated with cats. "For four thousand years," writes another art curator, Nora Scott at the Metropolitan Museum of Art in New York, "the sacred enclosure at Bastet's shrine was probably never without an occupant in whom the spirit of the goddess could manifest itself."[13]

In 1958 the Metropolitan Museum in New York proudly announced its acquisition of the Cat of Bastet, a fifteen-inch bronze statue that, like other such items (the museum said), would have housed a sacred cat mummy, or perhaps a cat belonging to a person wealthy enough to give it a special burial.[14] Museumgoers fell in love with the cat statue and wore a path to her side.

In 1981 the statue was removed from display. As far as I have been able to discover, no public explanation was given at the time. Six years later a reporter visited the conservation lab at the Met for a completely unrelated matter. He noticed the Cat on a lab table, and heard a conservator remark about its "fall from grace." Evincing a cat's curiosity, the reporter pressed for more information; before long, museum officials admitted that they had for years harbored suspicions about the Cat's authenticity. The story soon broke in the *New York Times*: "Met Says Its Popular Cat Is Probably Fake." X-rays, it turns out, had pointed to features of the statue's bronzing that made a modern origin highly likely, rather than an ancient one.[15]

This famed object remains under suspicion, but the fact that some cats *were* kept in sacred coffins in ancient Egypt is beyond question. The Met's own website describes another cat coffin of Bastet, acquired two years before the controversial one.[16] Historians note a chronology that begins with a fascination with felines, dating back at least 4,500 years, when an artist decorated a tomb with a relief of a wildcat stalking marsh birds. Then came an association of cats with deities, and later the coffins and cat mummies. Originally cats were mummified so that they themselves would enjoy an afterlife. Later they were killed at young ages and offered up to the gods at temples.

Does the Met suffer from some sort of cat-related curse? The Cat of Bastet fakery was followed by another surprise in 2002. That year, one of its cat mummies was CT-scanned by the Albany Institute of History and Art. Scientists discovered a *dog* inside! The Institute notes, rather dryly, the Egyptian practice of mummy-faking. Small dogs, bones, or even mud were sometimes used as offerings to Bastet in place of cats.[17]

Egyptian cosmology was bloated with animals. Joining the smaller cats were outsize ones: lions, or part-lions, dominate in many sacred contexts. The Sphinx, after all, is a "binary spirit-creature" with a lion's body and a sun-king's head.[18] Dogs, falcons, bulls, cows, crocodiles, baboons, and cows were buried in "vast necropoli of subterranean galleries." In preparing the animals for interment in these sacred cemeteries, people took great care. They embalmed the animals, anointed them with oils and perfumes, and then wrapped them in linen before burial.

The sacred status of animals was not appreciated by all ancient peoples. In the fifth century CE, when they discovered evidence of Egyptian animal worship, Christians set ablaze some of these cemeteries. The conflagrations destroyed the sites, and attested to the great backlash against pagan ritual that dominated this time period.[19]

That baboons were among the animals prepared ritually for burial in Egypt is fascinating news for a primatologist like me. Ancient priests worshipped Thoth, the god of writing and magic (the "scribe god"), in the form of baboons (and also ibis). A group of baboons lived in a compound of forty rooms and courtyards at a place called Tuna el-Gebel, south of Cairo. Today it's ecologically a desert but archaeologically a treasure trove. Documents offer glimpses of ancient life there, but most telling are the baboon mummies, arranged in vast numbers in a stunning catacomb.

I love the idea of the baboon representing the god of writing. In my study of Kenya baboons, I never once saw any scribelike tendencies on the part of the monkeys. What I did see were communal animals who were all about communicating with each other, day by day, season by season, year-round. Barking alarm calls to announce lion sightings; quieter vocalizations and come-hither faces to welcome

new babies; eyelid flashes and ground-slaps to warn a rival of serious-
ness of intent. I can only wonder at exactly how the god-of-writing
and baboon-symbol linkage came about in ancient Eygpt, but it makes
an intuitive sense to me.

So many animal-laced mysteries haunt our understanding of the
ancient world. Ancient Egypt's art and artifacts, starting in the Old
Kingdom's pyramid-building period and continuing past the birth of
Christ into the Greco-Roman period, illuminate for us the power of
animals to move the imagination. Even as the animals used, and the
meanings created, changed from place to place and period to period,
at base was a confluence of the animal, the human, and the divine.
Sometimes the burials and mummies allow us to reconstruct something of the rituals the Egyptians carried out. At other times the ratio
of known-to-unknown tilts and we are left only with a tantalizing
sense that ancient people brought animals into their religious lives. A
murky situation like this prevails at another ancient and religion-
steeped place: Stonehenge.

Unlike Catalhöyük in Turkey, or Tuna el-Gebel in Egypt,
Stonehenge in England is a household name, and a huge tourist draw.
Stonehenge's Sarsen Circle, with its upright sandstone blocks and
crosspiece lintels, is probably the monument's iconic image. Almost a
million visitors trek to southern England's Salisbury Plain each year,
to wonder at why humans erected these circles of massive stone, at
just about the same time the great pyramids were going up in Egypt.

Wonder the Stonehenge tourists well might, because scientists
have themselves come to no definitive understanding of the monument's meaning. Almost certainly religion is involved in some way,
and at the heart of a new hypothesis is a surprising source of information: cattle teeth.

It's tempting to think of the construction of Stonehenge as akin
to the building of a monument like, say, the Washington Monument
on the Mall in Washington, D.C. That ninety-ton towering structure
is reminiscent of the Old World simply because it is shaped like an

Egyptian obelisk. And like any monument worth its salt, it took a while to build: thirty-six years to be exact (although the Civil War and financial problems caused a long gap in construction).

By contrast, Stonehenge was remodeled and reshaped over 1,400 years. The process of comprehending the events of those years, and what the monument meant to the people who made it, goes on still. People were buried there, we know that much. Around 5,000 years ago, even before the massive stones were erected, cremated human remains were placed in the earth. No final resting place for the masses, the Stonehenge cemetery accommodated relatively few sets of human remains. Perhaps it was devoted to the dead of a ruling family over an extended period of time.

Recent skeletal evidence reveals that Stonehenge was also the final home for people with certain deformities, a clue that suggests ailing or injured pilgrims visited Stonehenge in hopes of being healed. Inscriptions in Wales, the origin point for some of the stones, bolster this suggestion by attesting to the stones' powers of magic. People, like the stones, may have come a far piece to reach Stonehenge.

Now we're getting closer to the role for cattle teeth in this story. Three kilometers from Stonehenge is a place called Durrington Walls, boasting a huge circular earthworks and a circle of timber. Most important, people *lived* at Durrington Walls. During midwinter and midsummer, people occupied homes there, anywhere between 300 and 1,000 of them. Other buildings, separate and without any household-related artifacts in them, may have housed priests, or formed the setting for ritual activities.

Notice anything intriguing about the timing here? People resided at Durrington Walls at two specific periods of the year: a perfect match with the summer and winter solstices. It's tempting to conclude that people used both sites—Durrington Walls and Stone-henge—to perform specific seasonal rituals. Notably, Stonehenge's avenue is aligned with the summer solstice sunrise, and the winter solstice sunset, whereas Durrington Walls' avenue lines up with the summer solstice sunset and its circle of timber with the winter solstice sunrise.

Taking a longer view, the chronology suggests that the Durrington Walls people may have built Stonehenge's earliest stone circles.

Finally, the cattle teeth! Surpassing even the intrigues of the buildings at Durrington Walls are the six teeth found in trash deposits. The coolest thing about them is not their structure, but their chemical signatures. Here's why: soil varies in its chemical makeup from place to place, and when young mammals' teeth are in the process of forming, the local chemicals enter and become fixed in the dental material. Analysis by scientist Sarah Viner of the ratio of two chemicals (varying forms, or isotopes, of the element strontium) in the cattle teeth shows that these animals were raised in areas of Britain far from Stonehenge—perhaps as far distant as Scotland and Wales. Did the cattle accompany pilgrims to Stonehenge? If so, were the animals themselves involved in some kind of religious ritual? If not, why would they be herded such a long distance? These questions tantalize, and no answers are forthcoming.

Archaeologist Parker Pearson views Stonehenge and Durrington Walls as connected parts of a ritual landscape. At Stonehenge there would have been a twin focus on healing and on veneration of the dead. If he's right, we have evidence for a major religious center in prehistory, one where animals played a role—but the nature of that role is shrouded in lost meanings of the past.[20]

THE LION

Particularly resonant in the religious imagery, stretching right back in time to ancient Eygpt, is the lion. Given the Egyptians' focus on wild and domestic small cats, it's unsurprising to learn that this biggest and fiercest of all felines held a sacred place in their culture. Inscriptions associated with the pharaohs attest to the lions' sacred status. The Great Sphinx at the Giza pyramids, carved out of limestone bedrock, is an image instantly recognizable to people the world over, and the world's first monumental sculpture. One part of a great funerary complex, it portrays a man's head (almost certainly that of an Egyptian king) and a lion's body.

Nothing, however, beats finding the real animal in the flesh—or rather in bone. Archaeologists did just that in the tomb of a woman called Maia, located at Saqqara.

Maia had been the wet nurse to the famous Boy King, Tutankhamen. She was entombed when she died, over 3,000 years ago. In subsequent years the same tomb was reused for burials of humans and cats, and among the cats archaeologists discovered an unexpectedly outsized example: a lion that had died of natural causes. Given its presence among "catacombs connected to the cult of animals," scientists concluded that the lion, too, was no ordinary animal, but some representative of the divine.[21]

In this veneration, the Egyptians were not alone. Sacred lions graced a wide swath of the ancient world from Persia to Babylonia. The Lion of Judah represented the Israelite tribe of Judah, then came to denote, more broadly, protection and strength in Judaism. The Jamaican Rastafari religious movement, perhaps best known through the music of Bob Marley, rests on the belief that the Ethiopian emperor Haile Selassie (1892–1975) was God Incarnate. Selassie, who, Rastafarians believe, was a direct descendant of the tribe of Judah, took at his coronation the title "Lion of Judah." The lion is among those animals centrally linked to the Buddha as well. Indeed, the Buddha's teachings are explicitly called "the Lion's Roar."

In seeking to discover whether the lion was traditionally associated with Islam (I found no such linkage), I stumbled across a fascinating video. Posted on the Internet, this three-and-a-half-minute clip was filmed at a zoo in Baku, Azerbaijan. The captions proclaim that the lion in question vocalizes the name of Allah. As I began to watch, I noted the lion lying still in an enclosure of bare concrete and iron bars. The lion soon stands, and begins to vocalize in short rhythmic bursts. It is clear that onlookers perceive something important to be happening because they begin to vocalize back; not a speaker of Arabic, I am at a loss to translate the specifics, but I recognized a call-and-response interaction between lion and people. (Keep in mind it's the *people* and not the lion who do the responding.)

More than a hundred comments on the video have been posted, and include overtly emotional responses like this one: *I was shocked when I saw the lion saying Allah but I always knew animals do pray. This is a good way for Muslims to really open their minds and hearts and to start praying.*[22] The video evoked a sense of pride in some viewers. *It's a miracle and I am proud to be a Muslim,* wrote one person; *It brings tears to my eyes for being so lucky to be born as a Muslim,* wrote another.

Nowhere is the lion more critical in symbology than in Christianity. The lion is ubiquitous in Christian scripture and in art. To think of the birth of Jesus usually brings to mind a nativity scene, with the ox and the lamb in the manger, accompanied perhaps by the trio of camels that arrived with the Three Wise Men. Yet as Laura Hopgood-Oster points out in her book *Holy Dogs and Asses: Animals in the Christian Tradition,* ancient stories connect the infant Jesus with lions. One tells of Jesus comforting his mother, who feared the beasts trailing them in the desert; the lion, Jesus says, has come to serve rather than to harm them.[23]

No ordinary person can tame a wild lion, a beast with inhuman strength that wields muscles, claws, and teeth to back up its reverberating roar. What better animal to symbolize the special qualities of Jesus (and some of his followers)? But the lion itself may be sensitive and astute. In another story, a young Jesus left Jericho along a popular road, and passed by a lion's den. People nearby were filled with fear, but Jesus entered the den and walked among the great creatures and their cubs. And, we learn, the lions worshipped him.

The lions triumphed where the humans failed, for it was they who recognized Jesus for who he really was. Outside the lion cave, Jesus said to the people, "How much better are the beasts than you, seeing that they recognize their Lord and glorify him; while you men, who have been made in the image and likeness of God, do not know him!"[24] Jesus then travels along some ways with the lions, and commands them to hurt no one—just as he commands the people not to hurt the lions.

In this story is raised the specter of the *imago Dei*, the great separation between humans and animals insisted on by the Christian doctrine that asserts only humans are made in God's image. This doctrine hints at the complex treatment of animals in the Bible.[25] As Hopgood-Oster notes, animals are sometimes blessed and cared for in the books of the Bible—they had a great Ark built for them, after all! Other times they are held apart or subjected to human domination. In this regard, the Christian lion is especially interesting for its role in events where kindness is expressed mutually between animal and human.

Think of the story of Daniel in the lion's den. The victim of a plot against him, and sealed by King Darius into a lion's den, Daniel nonetheless survives an entire night among the fierce beasts. The next morning he explains that God sealed the lions' mouths, and thus protected him. The lions go on to savage Daniel's enemies (and their families).

Sometimes individual lions forge an emotional bond with humans, rather than being mere vehicles of God's action. The apostle Paul, in one account, converses with a lion, and the lion requests to be baptized. And Paul does it—he baptizes the lion! The image of Paul carrying out this most sacred of Christian rituals, the very act that anoints a person as a Christian, on a nonhuman being is immensely moving. It makes me wonder: What would a theology of the future look like if it were rooted in the principles of this story? What if animals were embraced into a circle of religiosity, without regard to species and without insistence on setting humans apart as spiritually superior? I'll have more to say about this vision of theology in a later chapter.

Like Daniel, Paul, too, was thrown to the lions, and like Daniel he survived. In fact, he reencountered the very lion who he had earlier baptized![26] Whatever our relationship to the Bible, we can recognize that there is something powerful being told here in allegory and symbol, another linkage of the human, the animal, and the divine, just as we found in ancient archaeology. Another example involves Jerome, a fifth-century saint, who was immortalized in paintings by artists as famous as Bellini, Giotto, and van Eyck. Lesser-known painters like Zanetto Buggatto, Niccolo Antonio Colantonio, and Vittore Carpaccio focused on Saint Jerome's interaction with a lion (the van Ecyk, I should note, does include a lion at the saint's feet).

Both Buggatto and Colantonio portray the saint and the lion alone together as Saint Jerome tends to the lion's paw, wounded by a thorn. Whether outdoors or in an indoor room graced with books, the scene painted by each of these artists radiates serenity, as Jerome and the lion together attend to the thorn. The Carpaccio painting differs radically. It captures the fear of others in the presence of the lion. Their bodies strain away from the beast, just as Jerome remains calm, attesting to the nature of his relationship with the lion.

What a surprise it has been, in writing this book, how often I spend a day immersed in reading about some animal species, only to discover that same evening a newspaper article or television documentary that adds depth to my understanding. With lions came the single most arresting example of this phenomenon. Hours after I wrote about Saint Jerome's embodied calm, I relaxed with our new HDTV. Remote in hand, I surfed the high-definition channels to look for a nature show or animal documentary. The screen filled with lions. An interesting-enough coincidence, I thought, with mild interest. And then I began to watch Dave Salmoni in action.

I had stumbled onto the Discovery Channel's *Into the Lion's Den.*[27] Canadian animal behaviorist Salmoni had traveled to a private game reserve in South Africa to test his idea that he could approach wild lions at close range, based on nuanced observation of their behavior and communication. Though he had prepared extensively (and scientifically) for this dangerous project, I knew none of this history at the time. I just knew I was absolutely riveted.

During my baboon-watching year in Kenya, toxic snakes unnerved me by day, but it was lions that filled my dreams at night. More nightmarish than anything, the dreams continued even once I had returned to the States. There I'd be, at home and relaxed in my dream life, only to discover with heart-pounding shock that a lion had been hiding behind the couch in my living room. All the stresses of graduate school seemed to concentrate themselves in tawny maned bodies that popped up to scare me.

But here was Salmoni, in real life, walking out in the bush with just a stick and all his nerve. Over and over again, he uttered the same phrase in a low-toned, singsong cadence: "Big lions." In this nonthreatening voice, he conveyed that yes, he was there and moving closer to the pride, but he had no intent to harm. His body movements became a perfect mirror of his voice. With exquisite care he moved, turning his body or sitting down on the grass with a slow and smooth trajectory.

Dave's cardinal rule was to back off immediately, but calmly, when a lion growled or showed any sign that it had crossed from alertness to aggression, or even to a mild "go away" signal. The lions, then, remained, and presumably felt, in control. I appreciated this; it was not only smart in terms of Salmoni's survival, but it meant the lions were not stressed.

Salmoni's acuity was evident when he "read" a lion's tail-twitch, pace, or stare. Gradually, over some weeks, he edged closer to the same small pride, a group led by a large, elegantly maned male. A pride with cubs, no less! That he closed in to seven yards astonished me. Salmoni's colleagues parked themselves in a jeep nearby in case a need arose for rescue, but when he was close to the lions, Salmoni was on his own.

Salmoni is not, of course, a saint like Jerome, nor did he evoke any sort of spiritual relationship with the lions in the documentary. His knowledge was based on acute training and insight, based in science. What I found so remarkable was the *mutual* communication between the lions and Salmoni. The adult lions, male and females both, "read" Dave's body language just as he read theirs. It radiated what I had noted in the paintings of Saint Jerome—a quality of stillness.

On the TV screen, the latent strength of the lions was visible. Like a geyser ready to blow or a volcano about to erupt, but with the will to control their own behavior, the lions emanated power ready to unleash. No one watching could forget that a charge could come at any moment. The extreme sight of an undefended human being so close to wild lions brought home to me just why the lion is an ideal symbol for the power of God, for any person or people or tribe closely connected with God, or for a sacred force of any nature. If *this* relationship can be forged up close—if Jesus, Saint Paul, Daniel, and Saint Jerome can walk among and touch lions, so that the lions relate with them—the sacred nature of the interaction is communicated to all.

We may note a parallel between the biblical stories and a famous tale of Aesop's. The escaped slave Androcles enters a cave only to discover it is already inhabited by a lion. The lion lifts up his swollen and painful paw; spotting a thorn embedded there, Androcles removes it and tends the wound. Later, captured and thrown to the beasts, Androcles meets the very same lion again. The lion recognizes him, and adores him. The outcome of Androcles's story is profound: both he and the lion were freed. Through mutual kindness came liberty.

Lions, secular and sacred, are everywhere: the New York Public Library, the Art Institute of Chicago, innumerable sacred buildings. The next time I go cathedral-hopping in Italy, it's lions I'll be looking for. Hopgood-Oster says that lions grace the façade of the basilica in Assisi, the cathedral where Saint Francis grew up; they anchor the pulpit at Siena's great cathedral. The continuity of the lion across centuries and faiths is striking.

THE CLOISTERS

One chilly spring morning I stood in a medieval world: the Cloisters at Fort Tryon Park in upper Manhattan, an outpost of the Metropolitan Museum of Art, just across the Hudson from New Jersey. Art and architecture from the ninth through fifteenth centuries are housed at the Cloisters, imported from monasteries in France. There, among the elegant architecture, the golds and reds of the intricate tapestries, and the variety of household objects that brought the past to life, I felt enveloped in animal images fierce, fanciful, and fantastic. The Arch with Fantastic Beasts was probably my favorite of all.

Made of marble, originally from a twelfth-century French church, the arch depicts eight creatures in a carved parade. Five of the animals are species hybrids: a manticore, with a lion's body, a scorpion's tail, and a man's face; a basilisk, a cock-scorpion mix; a part-eagle, part-lion griffin; a bird-woman creature called a harpy; and a part-horse, part-human centaur. A dragon of familiar myth, a pelican, and (of course!) a lion complete the Arch's roster.

We already recognize the lion as a Christian symbol. The pelican spills her own blood in order to nurture her young, thus representing Christ's death and resurrection. But all the creatures, mythical and real, "had their specific lessons to impart" to people in medieval France.[28] In that period in Europe, bestiaries were hugely popular volumes that "revealed God's scheme and [how] animals often embodied, literally, the divine purpose,"[29] and the arch is an architectural bestiary of sorts.

In a nearby room at the Cloisters, dimly lit by a weak sun on the day I visited, hang the famous Unicorn Tapestries. Dating to around 1500, and French in origin, these glorious cloth panels seem to tell a compelling story of the capture, killing, and rebirth of the unicorn. A sequence of seven tapestries was meant to tell (or so scholars believe) a coherent tale.

The tapestries are vast, so much so that the eye struggles to take in all the details as the creature, single-horned and snow-colored, meets its fate. One startling panel shows the unicorn, surrounded by spear-wielding people on all sides, head down and feet bucking out. A viewer feels a strong sense of a coming bad end, and indeed the animal's death ensues. But the unicorn is no ordinary animal, and this one comes back to life. Is this art meant to represent the life and resurrection of Christ? It's an obvious analogy to suggest, though not an airtight one; the unicorn's resurrection lands him in a tightly enclosed pen. Captivity, not joyful freedom, is the tapestries' lasting image.

To discern what animals may have meant in a certain religious context in a day far distant from ours is to play a game of interpretation. Though limited only to Western art, *Nature and Its Symbols* is a useful guidebook to animal images, and plant images, too.[30] I'm not convinced that there's much mutual relating between humans and plants, but the symbology is fun nonetheless. The walnut, the book says, with its harboring shell and three part-structure, symbolizes not only protection and fertility but also the Trinity. Even if that's a heavy load for a simple nut to carry, the association makes logical sense. But who, other than a botanist, might have guessed that the cucumber portends human sin? This association came about, I learned, because the cucumber reproduces very quickly. The plasticity of these symbols is fascinating: when the Virgin Mary is associated with a cucumber, it shows that she *lacks* sin.

According to *Nature and Its Symbols*, the unicorn stands for purity and chastity. The lamb, historically an important sacrifice among Jews at Passover, in Christianity represents innocence, patience, and humility, and may stand in for Christ himself. Indeed, in Hubert and Jan van Eyck's painting *Adoration of the Mystical Lamb*, the lamb sits on an altar in the Garden of Eden, and sheds its blood into a chalice. Nearby are the crown of thorns and the cross. At the top is a dove representing the Holy Spirit, beaming its illumination upon the gathered multitudes.

Not all animal symbols point to venerable qualities such as

chastity and humility. The grasshopper, associated as is the locust with plagues and pests, may suggest calamity and death. In yet another painting of Saint Jerome, *The Penitent Saint Jerome* by Lorenzo Lotto, we find the expected lion, but it's in the background. As we gaze upon the central figure of the saint, depicted in the act of striking himself with a stone (in penitence), our eye is drawn to the tiny but prominently placed grasshopper. Here it embodies temptation, indeed the devil itself.

This same context-dependent meaning-making that we saw with the cucumber (does it mean sin, or freedom from sin?) applies as well to animal imagery. Biblical crows save human lives. It's a crow that Noah sends out (even before he sends a dove) to survey the world after the great flood. But even more often the crow heralds misfortune or even death. This complexity is a true reflection of the animal world. Indeed, I'll have more to say later on about the complexity of crows in particular: here is a bird that invites us to look through a portal into a world of mystery, where we don't always understand what we see.

MORE THAN SYMBOLS

Just as religious sites in ancient Turkey and Egypt brimmed with animals who felt pleasure and pain, who exulted and suffered, so religion scholar Laura Hopgood-Oster reminds us that in religion anywhere and everywhere, animals are more than symbols. The animals may themselves be sacred, able to "exhibit agency and play an active role in the unveiling of the holy."[31]

Of course, this idea makes no sense in relation to manticores, basilisks, and other creatures like those of the Cloister's Fantastic Arch. These animals of the human imagination never lived, suffered, and died. In any religious tradition, some animals may merely represent the sacred, but more often, flesh-and-blood animals help create it. Animals and people met at the very dawn of human religiosity. Animals participated in rituals, taught humans about how to live (and how not to live) through their own day-to-day example, and altered people's lives and moods through mutual relating.

Flesh-and-blood animals bring us back to this book's beginning, when a camel strode through the doors of the Cathedral of St. John the Divine on an October day's celebration of Saint Francis. And flesh-and-blood animals can take us even further into the anthropology of today's animal-human relating.

5

Animal Souls

•   •   •   •

CHARLOTTE A. CAVATICA (spider, to Wilbur the pig): Oh,
Wilbur . . . don't you know what you've already done? You
made me your friend and in doing so, you made a spider
beautiful to everyone in that barn. . . .

—*Charlotte's Web* (movie version, 1968)

IN ANCIENT INDIA LIVED followers of a sect called Jainism. The
core of Jaina belief and right living centered on empathy with all
living creatures. Extreme care was taken to avoid harming even the
tiniest insect. No spider's web could be brushed aside, no quick
motion could be made by a passerby that would startle the cow in the
field, and certainly no animal could be killed for food. Indeed, even
inanimate objects were thought to suffer.

Jains, writes Karen Armstrong, "had to move with consummate
caution lest they inadvertently squash an insect or trample on a blade
of grass. They were required to lay down objects with care, and were
forbidden to move around in the darkness, when it would be easy to
damage another precious creature. They could not even pluck fruit
from a tree, but had to wait until it had fallen to the ground of its own
accord. . . . All living creatures should help one another. They must

approach every single human being, animal, plant, insect, or pebble with friendship, goodwill, patience, and gentleness."[1]

For Jains, the life force is everywhere, and as a result nothing was more important, hour by hour, moment by moment, than empathy. A man named Makkhali Gosala was one of the great Jaina religious teachers. He lived about six centuries before Jesus (one set of accepted dates for his life is 599–527 BCE). Called by his followers *Mahavira* or "Great Hero," he preached his first sermon near a city in India called Champa. There, says Armstrong, "gods, humans, and beasts assembled to listen to the preaching . . . and formed a single, loving community."[2]

What's so striking here is the evident sense of mutuality: humans and animals interact, and affect each other, in Jaina belief and practice. From that ancient sermon through to the present runs a strong continuity of empathy in animal-human relating within Jainism. Today, millions of people in India and elsewhere practice Jainism; nonviolence (*ahimsa*) and empathy toward all creatures is a constant guiding principle. With Jainism, we'll embark on a journey that brings together animals, evolution, and religion in cultures around the world.

Animal hospitals called *pinjrapole* exist in India. Some are devoted to a single animal, as in a well-known *pinjrapole* located in Madras that houses sick, neglected, or old cows. At times, more than 2,000 cows are in residence. When animal hospitals are run by followers of Jainism, they may not be your run-of-the-mill animal-care facilities! In these places, reports religion scholar Christopher Chapple, employees who sweep up may extract the tiny insects that inevitably become caught up in the broom, and place them in a dedicated room where they may live unmolested. (Chapple notes that the Jaina brand of nonviolence sometimes results, paradoxically, in animal suffering. Animals who are old, sick, or injured, must be helped, but Jains reject euthanasia, even in thoroughly hopeless cases where an animal's pain would be relieved.)[3]

By most people's standards, the Jains are extremely restricted in the types of practices that they are allowed to carry out.[4]

Jains

- are vegetarian
- don't use cloth whose production hurts animals or humans
- take care to preserve life in everything they do

Jains are also not allowed to work at jobs that cause harm, for example:

- those involving furnaces or fires
- those in which trees are cut
- those involving fermentation
- trading in meat products, honey, or eggs
- trading in silk, leather, etc.
- selling pesticides
- selling weapons
- digging
- circus work involving animals
- zoo work

Some people would agree, whereas others would fervently disagree, that "zoo work" necessarily entails harm to animals. In any case, the actions of animal-human relating inherent in—or forbidden by—Jainism may be extreme, but the basic ideals behind them are found across religions. Just as the Mahavira addressed a mixed crowd of animals and humans in his first sermon, it is thought that the Buddha's first sermon occurred at a deer park.

The Buddha, a man named Siddhartha Gautama, also lived in India, at a time period (around 563–483 BCE) that probably overlapped with that of the Mahavira. This is not the place to delve into the differences between Jainism and Buddhism; for our purposes it is enough to emphasize the common value placed by both religions on compassion toward all creatures, who are not only capable of suffering but do suffer as an inevitable consequence of living. Probably early Buddhists didn't take things as far as the early Jainists did, in terms of avoiding harm to animals; the Buddha ate meat, for example. What's important is that the Buddha directed "the emotion of love . . . to the four corners of the world, not omitting a single plant, animal, friend, or foe from this radius of sympathy." Further, for forty-five years he "tramped tirelessly through the cities and towns of the Ganges plain, bringing his teachings to gods, animals, men, and women."[5]

Once again we can find a degree of continuity from the past to the present. Contemporary Buddhist practices involve special consideration for animals. A principle called "dependent co-origination" (*pratitya-samutpada*) gets at the cosmic understanding behind this consideration: "Nothing exists in and of itself but only as a context of relations," as Brian Edward Brown explains it. Work for the environment, legislation to protect animals, and other steps to care for the world's "innumerable communities of nonhuman beings" are critically important but will only succeed, Buddhists say, to the degree that this principle of complete relatedness of all life is grasped.[6]

As we have seen, to draw connections between the past and the present in terms of religious thought and practice is tricky. It cannot be done as a straight line or a simple trajectory, because present actions are rarely if ever isomorphic with past ones. In the case of Buddhism, views of animals shift depending on time and place. Still, as would be expected, a thread of continuity can be found between the Buddha's original teachings and a special care for animals in the actions of Buddhist monks today. In a remote region of Thailand we can find a dramatic example: a group of monks has opened the grounds of its temple to monkeys, horses, deer, peacocks, geese, wild pigs—and tigers.

At the Pha Luang Ba Tua temple, monks walk huge tigers on leashes, and explain to interested outsiders that the mighty animals have adopted an attitude of nonviolence. (I find this claim fascinating, but I have no plans to test it with a personal visit.) The sincerity of the monks is beyond question. They are forbidden by their beliefs to kill other animals to provide food for the tigers, so the cats eat—and seem to thrive on—dog food. Each tiger is cared for by a monk; the pair is matched up based on the abbot's view of like personalities. If the monks' plans come to fruition, one day the tigers will move to a moated island where they will roam free, and never be caged as they are now for part of each day.[7]

Keeping in mind the location of the Buddha's first sermon, it's intriguing to reflect on the role of deer parks in countries with large Buddhist populations, like Japan. At Nara, Japan's ancient capital (a city once called Heijo), the free-ranging deer were considered for many centuries to be sacred beings, divine representatives of a god associated not with Buddhism but with the Shinto religion. (More recently, the deer were officially named by the Japanese government as national treasures, a secular rather than sacred designation.)

In the park are both Shinto and Buddhist temples; people who come to the park may be of any faith, or none. The deer roam at will, and people who wish to walk with them or feed them are encouraged to do so. Visitors at Nara may also purchase small aquatic animals and release them in a ceremony of liberation. Small children are encouraged to place goldfish in the pond and feel the exhilaration inherent in sponsoring the freedom of another living creature.

In the Christian tradition, one need look no further than the famous figure of Saint Francis, honoree of the much-celebrated "blessing of the animals" ceremony in October, to find a parallel. Take a moment, if you can, and gaze at Giotto's painting *Saint Francis Preaching to the Birds.* If you can't transport yourself to the Church of St. Francis in Assisi, Italy, an art history book will do nicely, or simply seek out the image on the Internet.[8] Here we see the haloed Francis, gesturing gently but intently toward a number of birds, who in turn watch him as he offers them the word of God.

Born in Assisi in the twelfth century, Saint Francis embraced poverty and devoted his life to God. A man with a fascinating life history (Franco Zeffirelli's movie *Brother Moon, Sister Sun* offers one version), he is best remembered today as the patron saint of animals.[9] But what was Giotto trying to get across, with the image of Francis preaching to the birds?

The painting commemorates what published stories describe as a real event: one day Saint Francis noticed a number of birds in trees by the roadside, and stopped to preach to them. The birds flew closer to him. Addressing them, the saint said, "My sisters the birds, much are ye beholden unto God your creator, and always and in every place ought ye to praise Him, because He hath given you liberty to fly wheresoever ye will, and hath clothed you on with twofold and three-fold raiment. . . . Guard yourselves, therefore, my sisters the birds, from the sin of ingratitude and be ye ever mindful to give praise to God."

The painting beautifully captures the mutuality of the moment: Saint Francis relates to the birds in an emotional way, and they relate to him in kind. As the story goes, the response of the birds to the preaching was immediate. With opened beaks, stretched necks, and spread wings, the birds reacted to the words they heard. Indeed, the preaching saint and the enthralled birds rejoiced together, in shared devotion. When, at his sermon's conclusion, Saint Francis made the sign of the cross, the birds rose singing into the air, and flew off in four directions. Thus they would carry the message of Saint Francis to all corners of the globe.[10]

As with Jainism and Buddhism, we can recognize in some current Christian practices the roots of old ways. More and more, Christian theology embraces animals and animal issues. The world's first academic fellowship in theology and animal welfare was established at Oxford University in England. In his role as Fellow, Andrew Linzey writes passionately about animals. In his words we recognize a direct link with the Buddhist and Jainist focus on suffering: "Christians whose eyes are fixed on the awfulness of crucifixion are in a special position to understand the awfulness of innocent suffering. The Cross of Christ is God's absolute identification with the weak, the powerless and the vulnerable, but most of all with unprotected, undefended, innocent suffering."[11]

From these examples, we can begin to see taking shape a fundamental fact: for *all* major religions, from those of the East, to

Christianity, and also to Islam and Native American cosmology, compassion and kindness to living creatures is a central ideal. This fact is a powerful launching pad from which to continue our exploration of the relationship between animal-human relating and religiosity.

Does being-with-animals today bring people closer not only to the earthly world but also to a supernatural one? Are animals "God's messengers," as the title of one book says? That is, does God (or, in other cultures, gods or spirits) speak to us and send us messages of love or comfort through animals, a message that can guide us toward greater compassion for each other? Does God himself have a relationship with animals? Do animals have souls? Where do agnostics and atheists fit into a framework of being with animals that entwines itself so intimately with religion?

BLESSING CEREMONIES

I have already described in the book's opening pages the majestic blessing ceremony at the Cathedral of St. John the Divine, in honor of Saint Francis. A similar ritual unfolds annually at the National Cathedral in Washington, D.C. In 2007 the most popular pet attendees seemed to be dogs. Watching a video clip of the event, an observer feels the blend of earnest emotion and lighthearted humor that permeates the ritual.[12]

A woman brings to the church her two Yorkshire terriers, adorned with matching pink bandannas. She explains, "I wanted to get my animals blessed. I'm Catholic, they're nondenominational." A woman speaker tells the assembled crowd, "Why do we bless animals? We do that because we recognize that God cares for animals primarily through us and that in turn the animals bless us and make our lives whole." With a sprinkle of holy water, a priest blesses a dog called Teddy, as his people look on: "Teddy, I bless you in the name of God, creator, redeemer, and sustainer, who loves you and your owner and wishes you many days of good companionship together." Teddy, it is clear, is less than thrilled with the feel of the water, and ducks between his owner's legs.

No camel is in sight, but a variety of species is in attendance. Holding a rat in one gloved hand, a child says, "I think that he's gonna be better and might not get as many cuts and stuff." A priest eyes a caged bird called Gaylord, and is warned not to put her finger in the bird's cage. She responds, "Yeah, Gaylord might think I'm kind of tasty. Oh, Gaylord, may God bless you and keep you. May God's face shine upon you and bring you peace, Gaylord."

What is the animal blessing ritual all about? As I see it, a blessing is offered in two different senses. The animals, as family members or dear companions, are brought directly into God's light in a recognition that God's face may shine upon other animals as well as humans. Equally the ritual provides a public way for people to acknowledge the gifts that animals bring into their own lives ("animals make our lives whole").

This assessment is fine as far as it goes, but does it go far enough? Are animals God's messengers? Do animals have souls?

GOD'S MESSENGERS

In "Religion Lite," animals are angels sent to us directly by God to make a difference in our lives. In their book *God's Messengers*, Allen and Linda Anderson say that the animal stories they offer "remind us that, yes, there is a God supporting each of us in whatever ways are best for our spiritual growth and renewal." One story tells us of a cat who disappeared for three days; the distraught owner began to pray for the first time in her life, and "felt the invisible hands of God—or perhaps the hands of loving angels" leading her back to her missing pet. The cat's disappearance led to the person's spiritual renewal. Another tale describes a dog named Fritz. As Fritz lay in the living room, he sighed. His human companion realized the dog was "right where I was, relaxed with God and at peace. . . . [He was] a meditating angel dog."[13]

God's Messengers was successful enough to spawn a cottage industry of books by the same authors: *Angel Cats, Angel Dogs,* and even *Angel Horses.* Entertaining and comforting, these tales may awaken or heighten our joy in being with animals. Their focus remains on what animals can do for us and for our personal connection with God—but what about a focus for animals?

A different approach is emerging within a cross-denominational movement that puts animal welfare front and center. An issue of *Best Friends* magazine, published by the sanctuary of the same name in Utah, includes a roundup of views across the faiths that brings home the power of this perspective.

Yusuf Saleem, an imam in a Washington, D.C., mosque, notes that the prophet Muhammad was compassionate to all beings, including animals. The "ark for animal compassion" is a fundamental principle in the Qur'an and must be a pillar of the Islamic way of life. Clearly the imam refers to the crucial importance of daily practice rather than only sacred reflection.

This need for action is echoed by Robin Nafshi, rabbi from a synagogue in New Jersey. Rabbi Nafshi exhorts the Jewish community to import compassion for animals into its ritual practices. Jews should "call on kosher slaughterhouses to immediately put an end to the inhumane conditions in which animals are kept, transported and housed," and seek alternatives to the cruelty that sometimes accompanies Jewish ceremonies (as in the traditional killing of chickens before Yom Kippur). Most eloquently, the rabbi writes, "We read in Proverbs that a righteous person knows the soul of his animal."

In both the imam's and the rabbi's words we feel the need for people to shoulder responsibility for animals that cannot care for themselves, and, equally, to develop our humanity to its fullest. This connection of a person's "righteousness" with the treatment of animals is extremely powerful, and expressed elegantly by Steve Keplinger, an Episcopalian rector from Arizona. In *Best Friends* magazine, Reverend Keplinger reflects on human-animal relating:

I would dearly love to tell you as an Episcopal priest that it is the secular world that has caused this broken relationship between ourselves and the rest of creation. But the truth is that this is our fault—the religious leaders of today's world. We have not interpreted our sacred text for our present-day situation. We have not followed the example of our predecessors of identifying those most abused in our culture and then taking the necessary steps to change the way of thinking that causes that abuse. . . . It is time for us to hear not just the cry of God's people, but to hear the cry of all of God's creation. There is nothing more important we will ever do.[14]

When I wrote to Reverend Keplinger at St. David's, his church in Page, Arizona, he expanded for me on his message. In Reverend Keplinger's view, a Christian rethinking of God's relating to the universe can be effectively coupled with action on behalf of animals. This belief is woven through the very life of his parish in a number of ways. Congregation members embrace the environment around them by hiking together, recycling together, and caring together for abandoned and neglected animals both in Page and on the nearby Navajo Reservation. To people in the community, sacred sustenance, in the form of animal blessings, is offered right alongside bodily nourishment in the form of pet food for their animal companions.

Most strikingly, St. David's has created a new liturgical season, one that begins on the feast day of Saint Francis and continues until Advent, in the weeks before Christmas. It is called "the season of creation." One Sunday during this annual period, the liturgy begins with a "call to humanity to recognize our relationship with all of God's creation." The service is an eclectic mix of music, reading, and worship; its material ranges from the words of Hildegard of Bingen, a twelfth-century Christian mystic, to those of Edward Abbey and Walt Whitman, master writers of environmentalism. A Navajo origin legend and an Ojibway prayer are included as well.

The Earth-creation focus is driven by Reverend Keplinger's belief that "Once people truly begin to believe that God is in the universe and not separate from the universe, it becomes very difficult not to treat all animals and all of creation as sacred beings."[15] Here we see, in the Southwest with its fusion of Anglo, Native, and Hispanic cultures, a vibrant synergy of worldviews on relating with animals that leads to action in the world.

What happens among Episcopalians in Arizona relates intimately to what happens among other religious activists around the world. A good example comes from Wales. Shambo, a bull, lived among the Hindu community at the Skanda Vale temple in Wales, where he enjoyed sacred status—until the government issued a slaughter order and Shambo faced imminent death.

It is true that for Hindus, cows have sacred status. Like most religious precepts, however, this one isn't so simple. Hindus do not worship cows in the sense of praying to them, but reverence for cows, expressed as a fierce commitment to preserving their lives, *is* a core Hindu belief. Mahatma Gandhi famously wrote that cow protection is the single most important outward manifestation of Hinduism. Anyone may offer a cow a piece of fruit or bread for good luck as they go about their daily routine.

Yet twenty-first-century life doesn't make the path toward serious protection of cows and bulls an easy one. In India's city of Delhi, 13 million people and 40,000 cows vie for space, sometimes on the roads in the middle of traffic. Cowherds are deployed in the streets to transport cows to more appropriate, less densely populated areas for everyone's safety.[16]

Shambo was a special case. For one thing, he lived in a sacred space, the Hindu temple, under the watchful eye of monks; he was loved by a religious community, and no one there wanted him gone. Still, he was tested under the protocol of the United Kingdom's animal-disease policies. When the results came back positive for bovine tuberculosis, Shambo became a worldwide symbol. A clash was playing out between a Western government's rules and a Hindu community's deeply religious views about cow protection, and Shambo was right at the center of the conflict.

Events escalated rapidly. It was in May 2007 that the TB results were announced, followed by the "intention to slaughter" notice. The monks responded within twenty-four hours, campaigning in a highly public way to save Shambo; within a week, 4,000 people had signed an online petition to spare the bull's life. One of the monks' statements read, "We could no more allow the slaughter of Shambo than we could the killing of a human being. Ultimately, we will be willing to defend his life with our own."[17]

It's impossible not to think back to Jainism here. For many people, the sentiment expressed in the fight to save Shambo would seem as extreme as a person handicapping his own movements to avoid quashing a bug. Can all life really be equal? Does a spider's web deserve as much protection as a human home? Is a bull's life equivalent to that of a human being? Hindus around the world, joined by Buddhists, Muslims, and Christians, affirmed their position, installed a webcam to broadcast any action around Shambo, moment to moment, and planned to form a human chain around Shambo to prevent his being removed from the temple.

What followed between early May and late July of that year was a protracted series of legal maneuvers, with experts called in on both sides. At least one international animal expert said that the risk of infection to other animals or to people was virtually nil, especially if Shambo was kept isolated at the temple. Counterarguments stressed the possibility of the TB's spread to other people or animals. When one side spoke, the other countered, and the media hung on every word.

At the end of the day, the government won, and Shambo lost. He was slaughtered on July 26. The prayers and chants of supporters filled the air around Shambo in his last hours at his temple home. Monks, together with their supporters, chained the gate to the temple and refused to allow the first contingent of health officials and police to enter; essentially, the human-shield plan was put into effect. When twenty police officers returned with a warrant, there was little recourse. Shambo was led away, and his human protectors expressed bitterness at what they considered a desecration.

Should Shambo's life have been spared? How should we proceed when a religious belief on behalf of animals collides with a potential threat to public health? These questions evade easy answers. What is clear is that Shambo's situation, and his eventual death, stirred passions that many religious leaders around the world hope to tap in a systematic way.

In 2007, religious leaders met on Capitol Hill in Washington, D.C., to sign a proclamation of universal kindness to all animals. The proclamation's beauty is its call for specific action. Five paths to action are highlighted, for animals who are (1) homeless; (2) used in the food and fur industries; (3) abused in sports and entertainment; (4) used in medical research; and (5) living wild, in need of preservation and protection. The language of the proclamation is, in places, elegant: "We therefore invite all people of faith, under the guidance of their various faith traditions, to take up the mantle of compassion toward all of life and recognize that, as human beings, we are only part of God's creation and cannot presume to be all that matters in it."[18]

The religious feeling stirred by animals may run deep. Martin Buber, the nineteenth-century Jewish theologian, captures in his writing the emotional depths of mutual relating. For years I have cited passages from Buber's works in order to explain the emotional power that is unleashed when a group of people creates a sacred ritual, or when two people come together in relationship—or when an animal and a person connect. Many of my talks about apes and human evolution conclude with his statement, "In the beginning is the relation." Authentic relating for Buber comes only when two entities connect fully and openly with each other in the moment. When this happens, there is no sense of what the other individual can do for you, no attempt to examine or measure the various facets of the relationship, just full-on being together. This I-Thou relationship—or, more correctly, this I-Thou *moment*, because such intensity cannot be long sustained—is the deepest way to relate. Otherwise, one or both participants become an "it," that is, an individual seen in a detached or functional way.

Only when I read more widely, for this book, did I discover that one of Buber's early soul-stirring encounters came about in the presence of a horse. As a child visiting his grandfather's estate, Martin bonded with a mare. In a barn alone with the horse, he discovered the joys of total immersion with a nonhuman creature. In such a moment the world is shut out, and one's own mind is quieted. Life and love are poured into the space that person and animal create together.

One day, young Martin became caught up in thinking about how much fun he was having stroking the horse; he even focused on his own hand in the act of caressing the animal. In his book *The Souls of Animals*, minister Gary Kowalski explains Buber's own take on what happened next: "[In] that instant, he had ceased to relate to the mare as a friend and instead turned the animal into a thing: an object of gratification rather than a partner in pleasure. The horse also sensed the change. The next day when Martin returned to the stall at feeding time the horse no longer raised its head in greeting. Martin continued to pet the mare, but the relationship had changed."[19]

What Kowalski describes is a lapse in the boy's focus that led to a swift change from a beautiful I-Thou relationship to a more detached I-It one. There's no doubt in my mind that Buber's theology is brilliant, yet in reading about this particular example, I felt skeptical about the supposed fragility of animal-human relating. Do we really turn other creatures into "things" when we consciously reflect on our enjoyment of them? When my tiny tortoiseshell kitty Emma (known affectionately as "Emmy" or "Emmylou") stretches her small self out across my lap, and I lay a hand on her body to feel its purr, contentment wells up in me. Of all the many places Emmy could have chosen to relax and rest in our relatively spacious house, she chose my lap. As our two bodies—the small cat's and the larger human's—come into connection as a result of her choice, pleasure flows through us. By bringing that pleasure to my conscious awareness, does Emmy become any less Emmy-the-partner to me? Does she become a mere object? I don't believe so.

In a world where too many thousands of domestic cats are abandoned or abused, can the reduction of a creature to an object come about from a loving interaction such as mine with Emmy? Couldn't we just as readily posit that Emmy treats me as a thing because I am an object of comfort, with a warm lap to offer her? Or, more probably, that she considers me a place of warmth and yet is still capable of I-Thou moments with me?

Kowalski's description of what happened with Martin and his horse led me back to his original source, Maurice Friedman's biography of Buber's early years. The boy-and-horse account is there, and includes sentences that did not make their way into Kowalski's book: "Aware of how his relationship to the horse had changed, [Buber] felt himself judged. Later when he thought back on the incident, he no longer imagined that the horse had noticed his lapse into monologue."[20]

As adults, we no longer think in the same ways we did as children. Indeed, the horse is not likely to have intuited Buber's lapse into self-focus, or resented it so that the contour of the boy-horse relationship shifted. It's far more likely that the horse failed to lift his head in greeting because Martin gave off signals of guilt or unease, or because he was preoccupied by a belief that he had betrayed his bond with the animal. Animals pick up with exquisite sensitivity on small changes in our bodily posture and on the signals we send out about our state of mind. Intimate relating can be damaged by the subtleties of how we behave, to be sure—but we shouldn't fear that it's our own pleasure that dooms the process of genuine being with animals.

Friedman also relates aspects of Martin's early family life. When Martin's father, Carl Buber, "stood in the midst of the splendid herd of horses, he greeted one animal after the other, not merely in a friendly fashion but each one individually. When he drove through the ripening fields, he would halt the wagon, descend, and bend over the ears of corn again and again, finally breaking one and carefully tasting the kernels. . . . Carl Buber anticipated one of the most fundamental aspects of his son's later thoughts: that the man who practices immediacy does so in relation to nature just as much as to his fellowman . . . the 'I-Thou' relation to nature is a corollary of the 'interhuman.'"[21]

I would say rather than *anticipating* Martin's later view, Carl *helped grow* it. Where else but from mothers, fathers, and other loved and respected adults do young children come to experience that first joy of relating with an animal? This learning may take place at home and/or in partnership with an imam, a rabbi, or a reverend. Compassion for the natural world may come out of secular teaching or sacred teaching, but either way, the power of its force is evident. We can only wonder, would Martin Buber's I-Thou theology have had a different flavor had his early world not included communion with horses?

ANIMAL SOULS

With all this talk of animals and religion, isn't there an elephant in the room? There's one question I have skirted, though I dropped a hint by mentioning the title of Gary Kowalski's book. *Do* animals have souls?

In the Christian tradition, the answer would, throughout most of history, have come back as a thundering *No*. Christian doctrine traditionally claims that humans were made in the image of God, and other animals were not; humans have souls that survive their earthly death, and other animals do not. Of course, this position is an etymological irony, given that the very word *animal* contains the Latin word for soul (*anima*)!

To think historically is useful, but to stop in the past misses seismic changes in Christian thinking. Pope John Paul II invited Catholics to think hard on this issue. The first non-Italian Pope in over four centuries, John Paul II enjoyed an unusually large and devoted international audience. To my grandmother Feliksa Choroszkiewicz Klockowski, who emigrated from Poland to the United States as a teenager, and to millions of others, he was the beloved "Polish Pope." In 1990, the Polish Pope declared to an adoring global congregation that animals do indeed have souls. Yes, it's true that the Book of Genesis says that only humans were created in God's image. But the Bible also says clearly that God created humans by breathing life into them, and thus giving them living souls; certain biblical passages state just as clearly that animals have the breath of life. The Pope noted, "The existence therefore of all living creatures depends on the living spirit/breath of God that not only creates but also sustains and renews the face of the earth."

And then John Paul wove together these various threads into the fabric of a conclusion that stunned some and delighted others: "The animals possess a soul, and men must love and feel solidarity with our smaller brethren."[22]

Religious scholars might at this point insist on debating what a soul is, in the first place. Kowalski notes wryly that because of books like *Chicken Soup for the Soul*, the term is overused these days.[23] Is the notion of a soul necessarily tied to the sacred? Or to God specifically? If a soul *necessarily* implies God, what happens to the millions of people around the world who do not believe in a single God? They, as well as agnostics and atheists, are left out; under this stricture, they would probably be uncomfortable with the notion of an animal soul *or* a human soul. Yet Kowalski shows that a minister's definition may embrace all religious faiths—or none. While Kowalski at times links the animal soul to his own faith, as when he writes that "God is present in all creatures," he also discusses the soul in ways that resonate with anyone who relates keenly with animals: "Animals, like us, are living souls. They are not things. They are not objects. Neither are they human. Yet they mourn. They love. They dance. They suffer. They know the peaks and chasms of being."[24]

What a beautiful sentence: *They know the peaks and chasms of being.* This statement points to the unity of all creatures who make their way in the world. The rightness of Kowalski's words emerges from the nuance of his language: he writes not of *having* souls but of *being* souls. And, definitional debates aside, for thinking about animal-human relating, the Pope's declaration is crystal-clear: no thick line, no Rubicon, divides humans and other animals in terms of inherent value and worth. All creatures participate in and contribute to the world, and all are to be treated with kindness and compassion. In this, the Pope echoed Saint Francis. Indeed, it's no coincidence that in 1978, within weeks of his election as the 264th Pope, John Paul traveled to Assisi, Saint Francis's home.

Of course, what may seem forward-thinking and progressive—in some circles, shockingly so—within recent Roman Catholic tradition is quite limited from the perspective of other major religious traditions. Notice that the Pope stressed *living* creatures. A Buddhist might claim that such a declaration omits a large chunk of the universe!

In Buddhism, vitality is everywhere—as much in the rock that your feet stumble over as you walk your dog, as in the dog's soul or your own soul. Thich Nhat Hanh, the Zen Buddhist monk and author, writes about the continuity between those beings who are aware and think, and those who are unaware and incapable of thinking: "[I]t is impossible to distinguish between sentient and non-sentient beings. . . . Minerals have their own lives, too. In Buddhist monasteries, we chant, 'Both sentient and non-sentient beings will realize full enlightenment.' The First Precept is the practice of protecting all lives, including the lives of minerals."[25]

A cow, a frog, a monkey—all may reach enlightenment. There's so much fluidity in the world that animals may be reborn as humans and humans may be reborn as animals.[26] But this isn't some simplistic notion of your neighbor Joe dying of old age and coming back as the cat you see roaming around your backyard. It is more that each person may have been, in a past life and depending on specific circumstances and the outcomes of one's own actions, an animal. Once this fact is accepted, then the consequences of harming an animal come into sharp focus. A Buddhist teaching puts it this way: "When a person kills and eats any of these beings, he thereby slaughters my parents . . . he kills a body that was once my own."[27]

Is this an East-West divide, then? Do Eastern religions embrace the unity of all life, whereas Western religions tend toward an animal-human dichotomy, such that when animals are considered on a par with humans, it makes worldwide headline news as John Paul II did? As with every dichotomy we have considered so far, this one too simply divides up a complex world.

For one thing, it's a popular myth, but a myth nonetheless, that Buddhists inevitably embrace vegetarianism. Some sects do, others do not, and as we have already seen, the Buddha himself ate meat. This diversity doesn't mean, however, that killing of animals is taken lightly by any Buddhist. The first of five ethical precepts to which Buddhists vow to adhere is to abstain from the taking of life.

Consider what one Buddhist text says happens to people who have killed birds and deer with no regret: Consigned to a special hell, they "are forced to eat dung alive with flesh-eating worms as punishment for their misdeeds."[28] The notion of regret is key here. Indeed, a person's intent is given paramount importance in the Buddhist worldview. Any person may swat at an insect and inadvertently kill it. With a vehicle, any of us may kill a squirrel or bird, even though we try to avoid it. We may be given meat killed by others, or meat of an animal that did not die violently. As with any set of religious precepts, the principles are important but their expression is complicated by context, by time and place, by individual intention, and by the meanings different people or different sects invest in them.

Let's go further. The work of the American philosopher Charles Hartshorne shows how the continuum of life—the connection of sentient and nonsentient beings—may infuse Western thought. Hartshorne's view is often described as *panentheism*, a term that should not be confused with the more familiar *pantheism*. Whereas pantheism is the view that God *is* all things, panentheism is the idea that God is *in* all things.

Hartshorne's views are complex enough for this non-philosopher, at least, to require a good guide. The *Stanford Encyclopedia of Philosophy* describes Hartshorne's views this way: "All creaturely feelings, especially feelings of suffering, are included in the divine life. God is seen by Hartshorne as the mind or soul of the whole body of the natural world, though he thinks of God as distinguishable from the creatures."[29]

In other words, God, a separate entity, is in each person, but also in other entities in the world: bonobos and bison and blowflies, grass and trees and the sea. Hartshorne was a published ornithologist, in addition to an acclaimed philosopher. He was intrigued especially by birdsong, and concluded that birds enjoy singing. (Given Buber's horse and Hartshorne's birds, it might be fruitful for someone to research the being-with-animal experiences of leading figures who write about the continuity of all life.)

Though associated in the popular imagination with Buddhist or native American worldviews, ideas of the sacred as present everywhere in the world are not narrowly linked to any certain tradition of thought. The diversity of ways that sacred texts and practices embrace animals may be somewhat startling to those raised on the famous passage in Genesis that seems to give dominion over the Earth's creatures to humans: "And God said, Let us make man in our image, after our likeness: and let them have dominion over the fish of the sea, and over the fowl of the air, and over the cattle, and over all the earth, and over every creeping thing that creepeth upon the earth."[30] Christian scholars debate what "dominion" may mean, and as we have seen, the Bible is full of animal symbols that may be interpreted in a number of ways. Still, a historical tendency to interpret that passage as humans' right to *rule over* animals is undeniable.

Opening up to the ways in which animals are sacred in the world's religions offers us another way to link being with animals and developing our humanity. By no means the only path to compassionate practice with animals, religiosity is one vital path.

One recent fall, 9,000 of us gathered in San Diego for the annual meeting of the American Academy of Religion. The AAR, which celebrated its centennial in 2009, convenes professors, theologians, pastors, rabbis, imams, and anyone else interested in the serious study of religion. So legion were our numbers in San Diego that we spilled over from the city's massive conference center into several surrounding hotels.

One morning I tore myself reluctantly away from the warm outdoor air and a bayside breakfast in order to seek out a paper session in one of the hotels. It was organized by the Animals and Religion subgroup of the AAR, in itself another indication of the growing recognition of lived links between animals and religion around the world. This small gathering underscored for me what I was learning through my own study, the ways in which the history of religion is deeply intertwined with the history of animal-human relating. As Aaron Gross put it during the session, animal-human relating is "charged and holy," and religious figures, religious rituals, and religious scholarship through the ages reflect this fact.

6

Ravens, Shamans, and Dogs Who Dream

•   •   •   •

Gwylum, Thor, Hugin, Munin, Branwen, Bran, Gundulf,
Baldrick, Fleur: Names of ravens in recent residence at
the Tower of London. According to legend, at least
six ravens must be present at the Tower or the
Kingdom of England will fall.

SOUTHERN UTAH'S RED-ROCK WORLD is a desert-lover's oasis and a geologist's playground. Red-and-buff sandstone formations draw the eye as sculpted reminders of Earth's dynamic forces. Snakes, lizards, and birds go about their daily business, oblivious to the rock art on the sandstone around them, the petroglyph figures of animals and people drawn hundreds of years ago by the first Americans.

Yet animals aren't the only ones who may miss out on the wonders wrought by the other species around them.

Raptors, marsh birds, and songbirds flock to Utah, and bird-watching is a popular activity. This ornithological passion is found in many parts of the world. Millions of people devote themselves to identifying species by plumage and song, and to observing nesting and molting habits at various times through the year. The skills and knowledge of the most dedicated observers can reach astonishing levels, as people travel with binoculars, spotting scopes, and cameras from place to place in search of hard-to-find species. When a species is spotted in an unexpected place, ornithology enthusiasts telegraph the news to others through their local birding organizations and, increasingly, via the Internet. Once a golden warbler was spotted in a supermarket parking lot near the town of Kent in England, far out of its typical range. When it decided to stay local, and not fly off, 5,000 people showed up to take a look!

Once in a while an observer especially skilled or especially lucky catches a glimpse of a hidden world, a world unknown to humans, where birds—at least some birds—do far more than mate, reproduce, and parent the next generation.

In Utah one day, naturalist Craig Childs followed some ravens who flew into a canyon. Too numerous to count, the ravens "grumbled and cawed" in clear agitation at Childs's close approach. They mobbed him, and pelted him with pebbles.

Why all the fuss? No nest was nearby. Nothing at all, in fact, was visible that might explain the ravens' fierce protective response. Except, Childs noticed, a single stone that sat in the middle of the wash. Under the stone was a feather.

Childs left, but returned later to the canyon with his hiking companions. One of them identified the lone feather as an owl's, but there was more to see. Owl feathers were all over the place, each positioned with care, one atop another, or wedged into cracks in the rock. Childs climbed to a ledge and found "what looked like an altar, a stack of feathers, their quills shredded by raven pecks." When he picked up a feather, the ravens became agitated.

Ravens are complicated birds; that much we know. Their tendencies sometimes run to the murderous. They will kill other birds, but because they do not eat them, something other than a predatory instinct must be at work. Apparently ravens recognize *something* about death, because they gather in mobs when one of their own dies. But this feather-keeping is something else again. What could it be about? Had the ravens killed an owl, then somehow marked the event? Childs's thoughts ran wild: "Perhaps this is what we found in a remote, nameless canyon: the commemoration of a moral act. An owl had been defeated by a coalition of ravens. . . . They enshrined the dead owl's feathers, and gathered, perhaps every morning, to remember what had been done."[1]

I don't know if ravens gather in communal remembrances. I do believe that the owl feathers *meant something* to the ravens. I'm talking about a social meaning, an understanding that the feathers were worth protecting from human intruders, because they stood for something in the group's collective history.

To know more, a scientific approach is required. Does feather-caching go on in other raven groups? If so, are the feathers always of owls? How are the feathers arranged, and are they protected or visited after the initial caching? Individual ravens could be tagged and tracked as they approached owls, so that the process of feather-gathering and its aftermath could be watched in its entirety.

It's natural for a scientist to want rigorous confirmation of an anecdote about raven ritual tendencies. Vast numbers of people, though, would embrace without a second thought Childs's linkage of ravens with ritual behavior ("enshrining the feathers") usually reserved for humans—and would go much further. In a stunning variety of cultures, ancient and modern, the raven is divine, or a messenger of the divine.

In ancient Rome and the Celtic world, people interpreted the divine through the ravens' flight and speech. Tibetan culture embraced bird divination, and its texts offered interpretations that were amazingly specific in their nature: When a raven speaks in the northwest, it portends the coming replacement of a king. When it speaks in the northeast, a quarrel will occur. When the raven vocalizes *gha gha*, a person's wishes will be granted, but when it says *tat a*, a person will find clothing.[2]

In North American Indian groups, the raven appears in myth as both a god, perhaps even a creator god, *and* as deceiver or trickster. Sometimes, in Native America, the line between deity and messenger blurs, and Raven is considered to be both at once.

Why the raven? Why is *this bird* chosen, again and again across cultures, to deliver to humans some sacred message? Writing about raven augury, Eric Mortensen grounds his answer in both the reality of ravens and the supernatural nature of human thinking. "The lives of the real birds largely inform and determine the way they fit into the context of human religious perspectives," he remarks. People notice that ravens "are simply remarkably vocal and imposing long-lived" birds.[3] They pay attention to raven "speech" and to the birds' clever behavior. And people *really believe,* he says, that ravens can see into the future and report back to humans.

Around ravens there hovers, then, an aura of the unknown. A confluence of science and spirit suggests that certain animals may bring together myth and mystery in compellingly spiritual ways. Indeed, the raven opens up another window through which we can view animals and religion, and how they may come together in cultures beyond our own.

TOOY ALEXANDER

In the country of French Guiana, on the northern Atlantic coast of South America, lives Tooy Alexander. Tooy, originally from the neighboring country of Suriname, is a Saramaka, a descendant of escaped slaves who fled into the tropical rainforest after their forced arrival from Africa in the early eighteenth century.[4] Saramaka lifeways offer a rich mine to anthropologists who want to understand the blend of older African and newer South American traditions that mark this African-American culture.

Tooy himself is a healer. People come to him when they have arthritis, stomach problems, or love troubles. Clients, not only other Saramakas but also people from Brazil, Haiti, Martinique, even France, gather at his house. The scene is often crowded, because apprentices stand ready to carry out Tooy's instructions: they find and prepare vines, roots, leaves, and bark from the forest to aid the curing process.

Tooy acts as healer from his position within an intricate world, one characterized by fluidity among people, animals, and spirits. In the Saramaka worldview, a person's ancestors are vitally active in everyday life, spirit possession is common, and animals are threaded through a complex religiosity.

Tooy's life and Saramaka lifeways are illuminated by Richard Price. Rich, an anthropologist who has worked among the Saramakas for decades, is a colleague of mine at William and Mary. When I asked him to describe some of the ways in which Saramakas live with animals, he invoked an idea (already noted in earlier chapters) made famous by Claude Lévi-Strauss, that animals are good to think with. Rich explained:

> Saramakas live very closely with animals—as pets (including highly valued hunting dogs), as hunting quarry, and as supremely good-to-think-with (in metaphors, allegories, fables, folktales . . .). A few kinds of animals can possess humans and speak through them: *vodú* (boa constrictors), *watáwenú* (anacondas), *adátu* (a toadlike creature), vultures, jaguars, and so on. Animals that are sought in hunting can be prayed to: *pingo* (bush-hogs), in that case via the woman (the Mother of all Bush Hogs) who controls them, and others. There must be hundreds of songs about particular animals and hundreds of folktales.[5]

Tooy calls on the gods, and on his own ancestors, to deal with issues in his life and in the lives of his clients. In his own house, he lives side by side with nine gods, three female and six male. In his healing work, part of what Tooy does is to settle gods in people's heads. It's a kind of taming process, where the gods are coaxed to express their needs and desires through a human medium. It's not only the gods that are sacred; as Rich has explained, animals may be sacred, too.

At times, through the voice of a god, Tooy imparts lessons about the forest, about the howler monkeys and snakes and other beings who live there and who may directly interact with the gods. He explained once, for example, about the "great snake who lives at the base of hills," a snake who provides the river and sea gods with clothes "when they come ashore to walk among the humans."[6]

Despite his mentoring of the apprentices, Tooy sometimes laments the indifference of the younger generation. All this knowledge he has to share, about healing and sacred rites, about animals and ancestors, and who genuinely listens? "My greatest regret," he told Rich, "is that I have no one to shoot the breeze with, no one to pass things on to. Young people today just don't have the patience."[7]

Indeed, Tooy's religious practices cannot be learned in a mere year or two. He prays, drums, and sings in more than a dozen languages, even in addition to his native tongue and to the Creole language he adopts to communicate with others in his community. The Saramaka cosmology itself is an intricate linkage of the animal and divine worlds, with a multiplicity of gods whose forces must be controlled as much as humanly possible.

Even as we recognize its complexity, the Saramaka worldview may strike us as exotic. Most readers of this book probably think in terms of a single God (whether they believe in that God or reject him), not in terms of a panoply of gods living in our heads and our houses. For those readers, the ancestors may be important personally, as when great-grandparents' stories of immigration or struggle are told and retold as a way of remembering a family's past. It's less likely that the ancestors are involved intimately in people's day-to-day decision-making. Similarly, animals may join with us in religious ceremonies now and again, and certainly help anchor our emotional lives, but rarely are these animals thought to be locked in tandem with the gods in bringing about what happens on a daily basis.

Retreating to labels of exoticism is never enough, though, when it comes to comparative religion. Hard as it can be to grasp in a culture gripped by fierce debates over the very existence of a supernatural God, the larger world is full of gods and spirits! What may seem exotic to some, then, is just the normal state of affairs for millions: the living and the dead communicate; animals guide people into the spirit world; people transform into animals to go more deeply into that world.

No Rubicon divides monolithic religious systems from others as regards these three points, for that matter. Brian Morris remarks that the most famous spirit medium in all of history was Jesus of Nazareth. Jesus's visionary experiences are amply recorded in the New Testament: Jesus felt himself full of God the Father and the Holy Spirit, and grappled with mean-intentioned spirits, not to mention the devil himself. "Even the early Christian church under Paul," Morris writes, "seems to have all the hallmarks of a spirit-possession cult."[8] Jesus, Buddha, and other monumental spiritual figures are reported to have related intimately and spiritually with animals, as we have seen.

Now let's continue the process we started with Tooy Alexander in French Guiana, and broaden our vision of the ways people and animals around the world may be connected in spiritual realms.

SPIRITS AND SHAMANS

In shamanism, a spiritual master connects to the spirit world, very often through being with animals. A shaman's goal, simply put, is to help other people with whom they live: through a shaman's actions, illness may be put to flight, and bad fortune of other kinds may be reversed.

There is no universal right way to be a shaman; to imagine a set of instructions on "How to Be a Shaman" would fail to imagine all the variations and nuances in shamanic practices across cultures. Still, some core features can be identified. These come from anthropologists who have lived among hunter-gatherer foragers, a group of rapidly dwindling populations who live—or try in today's frenetic world to live—off the land. As their name implies, hunter-gatherers track game, collect highly nutritious protein matter from vegetables and nuts, and perhaps farm a little or trade with other groups around them.

Though they do not live in industrialized or consumer cultures, we must not think of hunter-gatherers as isolated throwbacks living in the midst of a forsaken forest or jungle somewhere. In Arctic Canada and Alaska, we may think of the Inuit (or Eskimo); in South America, the Ache of Paraguay; in Africa, the Hadza of Tanzania; in Asia, the Agta of the Philippines; and in Australia, the aboriginal peoples like those we met in the discussion of Australian rock art. These culturally and linguistically modern populations inhabit a non-industrialized world where ancestors and animal spirits thrive.

- In hunting and gathering groups, people's consciousness tends to be focused on the seasons, the stars, the weather, plants, and, of course, animals; nature is "pervasively animated with moral, mystical, and mythical significance."[9]
- The universe, for these foraging peoples, is multilayered: the regular living world is complemented by the spirit world. The present time exists because the initial conditions of the mythological past were transformed into what we live now and see now.
- Altered states of consciousness are key to the work of shamans; these are accomplished through drumming, singing, dancing, and reaching an extraordinary level of fatigue.
- Animals are central to shamanic practice, which often includes a "soul flight" wherein the shaman travels into other-than-here-and-now dimensions and may come into contact with quite dangerous spirits. In such cases, animals may act as protective spirit guides.
- Sometimes shamans may transform into such animals as bear, tiger, or jaguar. This process differs from one in which the shaman is helped along in the spirit world by an animal, because it is marked by personal transformation.

As can be imagined, the ability, indeed the willingness, to enter into trance states, to take soul flights, and to encounter dangerous spirits, does not inhere in just anyone. To become a shaman isn't a mere personal choice. Across cultures, people report that they became shamans after some strange and unexpected experience, a dream or vision that led them to feel disoriented or even crazy, left apart from society, and confused. Over and over again, in tales collected by the anthropologist Edith Turner, shamans explain that they felt terror or danger or impending madness as a result of what was happening to them. They didn't just feel "called" to be a shaman; it was a choice made for them by forces in the spirit world.[10]

This traditional shamanism, with its attendant challenges and terrors, differs starkly from what's called neo-shamanism, a phenomenon of popular culture today. In both traditional shamanism and neo-shamanism, animals may be involved as people relate to spirits, but beyond this superficial similarity the two sets of practices share virtually nothing.

Closely tied to New Age spirituality, neo-shamanism is open to all. Anyone may become a shaman with a relatively brief investment of time, effort, and, usually, money. For the price of a weekend's time and a hit to the checkbook, people may attend "shaman school" for forty-eight hours, drum, chant, and connect with their spirit animals toward the goal of making one's own life better. "The emphasis in neo-shamanism is essentially on personal empowerment," writes Brian Morris. The spirit-helpers and power animals used "tend to be conceived as rather benevolent or benign spirits."[11]

What tends to be absent in neo-shamanism are the qualities most essential to shamanism: the focus on serving others; the awareness of potential confrontation with not-so-warm-and-fuzzy spirits; and a period of being tested and having to learn the techniques involved. Edith Turner, after decades of studying religious practices around the world, emphasizes that social groups don't set out to plan a healer's initiation, as they might plan, say, a rite of passage for their sons and daughters who are just coming into puberty. Rather, there's that strange visionary experience that puts a person into contact with the spirits, an event so soul-shaking that it simply cannot be shrugged off. In the Arctic, Nepal, and Central Africa, the shamans who have been tapped in this way learn that they must work for good and not harm, but also, Turner writes, that they "must not attribute the power to themselves. This is the shaman ethic, and it is commonly found wherever Shamanism exists."[12]

Given the peculiar permutations that the term *shamanism* has taken on, some anthropologists prefer to avoid the term altogether. Tooy Alexander, the healer in French Guiana, helps people in specific ways that might fit a reasonable definition of shamanism. Richard Price does not use the word, however, when he writes about Tooy. The blurred applications of the term, including its current feel-good associations, are included among his reasons.[13]

Yet it's an undeniable fact that in some human groups, people communicate with spirits through animals. Sometimes shamans are involved, sometimes not.

THE REINDEER PEOPLE

In Northern Siberia, in tracts so vast and so cold that they connote in our language the farthest regions of human habitation ("The cabin where I'm staying is so remote I feel like I'm in Siberia!"), live the "reindeer people." The Eveny people of Russia survive in the coldest inhabited place on Earth, where ice reigns for three-quarters of every year and where the lowest temperature hits minus 96 degrees F. (minus 71 degrees C.). They manage this in large part because of their relationship with reindeer.

The anthropologist Piers Vitebsky traces the Eveny culture's focus on reindeer back thousands of years, to the time of domestication around 3,000 years ago. Ancestors of the Siberian Eveny lived in China. Through Mongolia and Manchuria, large standing stones marked graves or places of animal sacrifice. These stones were carved with animal images, predominantly reindeer. The reindeer were depicted in a sort of flight posture, and were, in some later incarnations, shown holding the sun or a person with the sun representing his head. Later on, when people's bodies were mummified in the Mongolia region, the shoulders were marked with tattoos bearing the same in-flight reindeer images as found on the standing stones.

Indeed, it was on the backs of living reindeer that the Eveny's ancestors migrated from China to Siberia. The linkage of reindeer and human transport is no myth: with the reindeer saddle and eventually the reindeer sled, vast new tracts of land in the taiga and the tundra could be traversed. True, dog sleds already existed by this time, but a reindeer's strength and pound-pulling ability exceed a dog's. Also, the reindeer's grazing habits made it a more convenient self-feeding transport animal than the dog, which must be provisioned with meat.[14]

The human-and-reindeer relationship goes far beyond the utilitarian for the Eveny people. When Vitebsky met his first Siberian reindeer, he was struck by the animal's *presence* in the world. I, in turn, was struck by the similarity with how it feels to gaze into the eyes of an ape. Other than the fact that I don't touch the apes I observe, I could have written this passage: "As I reached out to touch one of them I felt I was face to face with a creature of considerable presence. . . . Most compelling were the eyes, which were huge, soulful, and capable of engaging one with an intense gaze."[15] That "the eyes are windows to the soul" is a familiar enough saying, but how often is it applied to reindeer? For Americans who grew up with jaunty Christmas songs and Burl Ives's television specials, Rudolph the Red-nosed Reindeer is a more likely focus of thought than are real-life reindeer herds and reindeer's impact on people's emotional lives.

That impact comes across from listening to elders of Eveny communities, through anthropologist Vitebsky. Reindeer, say the elders, "were created by the sky god Hovki, not only to provide food and transport on earth, but also to lift the human soul up to the sun." As children, during the long sun-filled night each Midsummer's Day, the future elders became part of a community ritual to welcome the new year. They prayed to the sun for the good bounty of life, and prayed for their children, for their hunting, and for an increase in reindeer. Then "each person was said to be borne aloft on the back of a reindeer which carried its human passenger towards a land of happiness and plenty near the sun. There they received a blessing, salvation, and renewal."[16]

Exactly how this ritual from past days was carried out is not clear to us today; the details are lost in antiquity. But the resonance of the ritual for thinking about shamanism is inescapable. For Eveny people generally, the reindeer is linked to a kind of ecstatic flight. As might be expected, with Eveny people who practice as shamans, the situation is more complex. "Whereas laypersons could only fly on the back of a reindeer," Vitebsky notes, "shamans could turn into a flying reindeer."[17] Indeed, the word *shaman* originated among the Eveny and closely related peoples, and in these cultures the link among animals, humans, and spirits is extremely strong.

Is strong? I choose the present tense deliberately, because even today Eveny people explain the power of animals to affect their everyday lives. Reindeer have a will of their own. Most people, Vitebsky found, have a reindeer called a *kujjai*. If a person is in danger—even without awareness of that danger—the *kujjai* may choose to die instead, in a kind of selfless substitution process. Although dogs and horses may also die in place of a person, with these creatures, no personal will is involved. Only with reindeer does a death substitution come about because of the animals' own will. Reindeer alone make decisions to act on behalf of a person.

Eveny people dream about reindeer, and use the dreams' content to make sense of issues and problems in their lives, sometimes in retrospect. It's not that the reindeer offer solutions themselves; these dreams feature no cartoonish talking animals. Rather, people need to pay keen attention to how animals act, both in their waking and dreaming lives. One woman dreamed of an old reindeer. She wanted to ride him, but he could not move, and his neck was bleeding. A tear came out of the reindeer's eye, and the dreamer felt great empathy for the suffering animal. Precisely one year later, the woman's father became ill with cancer. She explains the dream now as prophetic.

Of course, it's not only reindeer to which the Eveny pay attention. One day a hunter awoke to find an eagle flapping its wings inside his tent, only to discover later that his brother had been dying at that exact moment. Animals die for people, and may also represent people who are going to die. When animal dreams or events of this nature are retold to others, a way opens up for people to express deep emotion in a publicly accepted form.[18]

Dreams can contain powerful messages, as anyone knows. But what happens when animals, too, pay attention to their dreams?

DOGS WHO DREAM

All of us who have lived with a dog know some version of this scene: A dog lies down for a nap, emitting a soft groan. She lies quietly. Then, with a shift of her body, she settles into a more comfortable position, and a light sleep. Her breathing deepens. Soon her body begins to twitch, and her legs jerk. She whines a little bit. We may indulge in a smile and a guessing game: Is she chasing rabbits in her sleep?

In the Upper Amazon region of Ecuador, dogs also dream. But there, the Runa people observe the dreaming process with great care. In this way, the Runa say, they learn what the dogs will do the following day. If, during sleep, a dog barks *hua hua hua*, this means the dog is dreaming of chasing an animal—and that the next day, the dog will enter the forest to pursue game. If instead the dog barks *cuai*, the next day will not be so routine. This vocalization foretells the dog's own death at the jaws of a jaguar.

Dog-human relating among the Runa is a complex affair. On the one hand, dogs are fed poorly, and often go hungry. On the other, they are imbued with fantastic abilities: under certain circumstances, they can understand human speech. And they most definitely have souls.

It's to the anthropologist Eduardo Kohn that I owe my understanding of the Runa-dog relationship. Kohn envisions an "anthropology of life," a discipline that ventures beyond a focus on human behavior to embrace the fundamental ways that humans' and other living beings' lives lace together. It's no wonder that fieldwork in Ecuador among the Runa led him to such a framework. For the Runa, Kohn notes, "all beings, and not just humans, engage with the world and with each other as selves—that is, as beings that have a point of view."[19]

Dogs, then, see themselves as persons: this is the Runa worldview. There's no claim here that dogs think of themselves as *human*. It's not that dogs concern themselves with the worries and joys of our human world. To grasp this point is to break the necessary linkage between "human" and "person," and replace it with one that instead yokes together "aware being" and "person."

In the twenty-first-century world of animal studies, a growing movement suggests that some animals deserve the term "person." Examples might include apes who use symbols to communicate with humans, or elephants who remember and rejoice at finding long-lost friends. (We'll meet apes and elephants like these in a later chapter.) Although scientists are right now keenly interested in dog cognition and communication, none to my knowledge has taken to thinking of dogs as persons. Why apes and elephants and not dogs? The difference seems to rest with *awareness*. In the view of Western science, apes and elephants are able to access their own reflective abilities and their own memories, but dogs are not.

The Runa, though, come at it from a different angle. Yes, species differ physically, but there's an underlying continuity in how different animals see and meet the world. Dogs, for instance, experience a vital inner life, and that life may be expressed in dreams. Because dogs and humans have souls, both dogs and people may come to an understanding of each other's experience of the world (what Kohn calls the "subjectivity" that comes with being sentient) by cautious use of certain techniques.

This transspecies communication is not without risk: the Runa want to *understand* dogs, not *become* dogs; the process must be controlled with great care. And here things become intriguing. The Runa can understand dog speech, but in order to bring about an equivalent understanding of human speech by the dogs, they, the Runa, must intervene. They must make the dogs into shamans. In order to accomplish this transformation, a person may give hallucinogens to a dog, and tie its jaws together, while instructing it via a kind of special ritual speech. Of paramount importance is that the dogs be prevented from vocalizing. "If dogs were to 'talk back,'" Kohn explains, "people would enter a canine subjectivity, and they would, therefore, lose their privileged status as humans. By tying dogs down, in effect, denying them their animal bodies, the Runa permit a human subjectivity to emerge."[20]

A clear asymmetry operates here. The greater comprehension, and the greater control, belongs to the humans. This asymmetry in the human-dog relationship is mirrored in the spirit-human relationship. Humans interact with the spirit world, the beings that live in the forest and are masters of the animals, much as dogs interact with humans. Whereas people are understood effortlessly by the spirit masters, for the humans themselves to grasp what those spirits are saying, they must ingest hallucinogens. Thus the Runa, in relating with spirits, and the dogs, in relating with people, are sentient, but in need of help to transcend certain boundaries (in one case the boundary is human-animal, and in the other, spirit-human). The important point is that when the boundaries are finally transcended, it is *souls* that come into contact.

As you might expect for people living in the forest as the Runa do, surrounded by wildlife ranging from monkeys and agoutis to peccaries and jaguars, it's more than a dog-human dynamic at work. People ingest the parts of other animals in order to gain insight into these creatures' consciousness. Especially if they have shamanic powers, individuals may take on "a kind of jaguar habitus" in life, and in death their souls may inhabit the bodies of jaguars. Even between nonhuman species, fluidity may prevail. Like jaguars, dogs are forest hunters, and dogs may take on jaguar qualities. Jaguars themselves, Kohn notes, are "the subservient dogs of the spirit beings" in the forest.[21]

At work with the Runa, then, is a powerful dynamic by which a vital life force is ensouled in animals. Animal-human-spirit boundaries blur in certain spiritual contexts where otherworldly forces are at play.

ANIMAL MYTHS AND TEACHINGS

Equally compelling, if differently manifested, aspects of the centrality of animals to religion come alive in other cultures. In Africa, according to the religious-studies scholar Kofi Opoku, there was, as part of the religious heritage, a conscious awareness "that the environment was not dead or inert," that animals "share the same environment, the same faculties, and the same experience of life and death" as people and so become a special "source of wisdom."

At first the generalization about "Africans" might seem an uncomfortable one. Is it sensible, we might wonder, to speak of "Africans" as a monolithic group when the continent is, and has been, such a source of cultural and linguistic diversity? Yet Opoku believes, in the case of animals and religiosity at least and in an echo of anthropologists' confidence in identifying aspects of hunter-gatherer cosmology, that a foundational worldview may be identified. This worldview can be summed up thus: "Humans are not the only beings with life, and everything that has life is potentially sacred."[22]

One striking arena where animals and religion come together in Africa is traditional origin myths. Before the world was the world, when it was only a wasteland of marsh, the Yoruba people say, divine beings sent a small party down from the heavens. This group included a hen meant to scatter soil across an area of the marsh, in order to seed the world. Next sent was a chameleon to inspect the hen's good work. Why a chameleon? Because it is such a deliberate beast.

By contrast, it was the hyena—anything but a deliberate creature—that was said by the Fajulu people of Sudan and the Madi people of Uganda to be responsible for splitting heaven and earth. The hyena bit off the rope that had once joined the worlds, and so birthed the world as we know it.

These mythic animals, Opuku notes, are agents of the Creator. They not only change the world, they create it in the first place. I love how the animals' particular temperaments are woven into the myths' unfolding. In similar fashion, the pangolin, because of its unusual nature, becomes a central figure of hope for a number of different African groups. Though lizardlike in appearance, the pangolin is in fact a mammal, one that births a single baby at a time and that may walk upright. When threatened, it rolls into a ball. Opuku remarks on the pangolin's "inscrutable incomprehensibility"; here is an animal that should be an impossibility! Because of its unique nature, the pangolin becomes in some African societies a symbol of what can happen against the odds, a symbol of hope.[23]

Continents away, we find that the buffalo's centrality to Plains cultures embraces the symbolic as well as the practical. As I'll discuss further in a bit, bison-related ritual behavior is a famous example of animal-based spirituality within Native America. In the cultures of Native America we find another cosmology in which religion, animals, and nature are understood to be linked fundamentally, in origins myths as well as moral teachings. To take a walk through the National Museum of the American Indian on the Mall at Washington's Smithsonian Institution is to see that link come alive.

First to catch my eye was an exhibit called *Our Universes: Traditional Knowledge Shapes Our World*, about (and created by) the Anishinaabe Indians of the Great Lakes and central Canada region. There, an animated film shows a buffalo talking with Anishinaabe people, and a bear teaching them how to fish. A quote inscribed on the

wall captures the theme of animals as teachers: "The animals that represent the Seven Teachings are spiritual animals. We can look on Earth
at the habits of the wolf or the bear and get our teachings and learn
from them. But when we talk about the wolf giving us humility, it's
not really the animal, the wolf. It's the spirit animal, the wolf."[24]

Here is the crucial distinction: it's not the four-footed, living-and-
breathing individual animal that gives wise counsel or leads to new
understanding, it's the spirit that's housed at that precise moment
inside the animal. The animal is a medium. For the Anishinaabe,
wisdom flows through the animal like river water flowing through to
the sea.

Continuing along further in the *Our Universes* exhibit, I learned
that seven key teachings are paramount also for the Pueblo people of
Santa Clara in New Mexico. Once again, each of these is represented
by a different animal. In this case, people associate each of the four
directions of the compass with a sacred mountain range, a color, and
an animal:

West: yellow, black bear
South: red, bobcat
East: white, badger
North: blue, lion

Information of this nature is compelling because at NMAI, the
exhibits are curated by native peoples themselves. Visitors learn, then,
about animals as linked with sacred mountains, or as the font of spiritual teachings, from the people who have lived those beliefs. A single
key conclusion shines through. As one scholar puts it, "within traditional Native cultures, animals can provide humans with an immediate and tangible way to communicate and identify with their
Creator."[25]

It's this very context, though, in which a blurring between traditional religious practices and neo-shamanic practices becomes worri-
some, because often it's an attempt to emulate Native America that

drives the weekend workshopper. The person who seeks personal transformation through a self-empowering version of shamanism may be wholly sincere. He may mean no disrespect (in fact it's more likely that he means to honor Indian cultures), but not all Indian people welcome this kind of emulation. Although neo-shamanism may be a source of economic empowerment for native people, it may carry more costs than benefits, as explained by the Indian writer and activist Vine DeLoria:

"The rapid expansion of the New Age [movement] has been unusually detrimental to traditional religions. Non-Indians can pay very attractive fees to Indian shamans, and there has been a good deal of pressure on traditional healers to spend their time working with non-Indians and neglecting their own communities." This pressure, DeLoria notes, has led to "an unusual number of Indian fakers who have invaded suburbia offering to perform ceremonies" for "anyone with the money to pay."[26]

People today are searching for ways to connect spiritually to the world around them. I understand DeLoria's dismay about those who seek an outlet of this type through neo-shamanism and thus negatively impact Native American communities—and I have no wish to promote that practice. Yet I also understand, I think, something about its source.

Many people feel a deep unease, a sense that something has gone wrong with our relationships with other animals in our industrialized, consumer-driven cultures. It's tempting to seek an antidote of some sort, and a ready channel may seem to be available via groups who are seen as purely "in tune with the Earth," and as blissfully reverent of animals. In this context, Native America becomes a natural magnet. It is misguided, however, to see in Native America a wholly virtuous pro-ecology culture. To believe that Indians never overhunted the buffalo is, for instance, to deny reality. Sometimes, in the past and in the present, Native Americans have chosen and do choose economic gain over the welfare of other animals or over environmental health generally.[27] The temptation to seek heroes (historically called "noble savages") may be strong, but Native Americans live with the same economic and political stresses as do we all. None of this complexity renders any less important the notion of animals as moral teachers in Native American culture.

From this departure point it may become easier to see a certain schizophrenia in the role afforded to animals in North American and European popular culture generally: we in present-day consumer cultures may bring animals into religious rituals, yet at the same time eat other creatures without a second thought. For the most part, we don't explain our cultures' origins by mythic reference to animals. We don't accord to bears and buffalo sanctified roles for teaching our children about our lifeways. Yet threaded through our cultures nonetheless is a bedrock acceptance of animals as moral teachers.

The moral role we give to animals becomes all the more striking the more you focus on it. And I do focus on it! Every newborn baby in my circle of family and friends receives a gift. It's a premature present, no doubt, given that tiny babies are not known for their ability to focus on details of character and plot. But, I can't wait to give it, and so I give Arnold Lobel's *Frog and Toad* books to new parents.

Books like *Frog and Toad Are Friends, Frog and Toad Together,* and *Days with Frog and Toad* tell the tale of two friends who stay together through thick and thin. These two amphibians—notably of different species—support each other. They share the joys of kite-flying together. When one is tired or discouraged, the other finds ways to cheer him.

Yet not all in life comes effortlessly, not even for frogs and toads, and hints of challenge permeate the books. One day, Toad visits Frog's house only to find a note on the door, addressed to Toad, announcing that Frog has gone out because he wants to be alone. Toad is taken aback, and wonders, "Frog has me for a friend. Why does he want to be alone?"

Through a garden, the woods, and a meadow, Toad searches for his friend, and finally spots him sitting on an island in the middle of the river. Fearing that his friend must be sad, Toad runs home and prepares sandwiches and a pitcher of iced tea, then returns to the river. He is anxious and worried that Frog no longer wishes to be his friend.

After some considerable effort, involving a swimming turtle who carries him and the food along in the water, Toad arrives on the island. He begins apologizing to Frog for all the "dumb" things he does and all the "silly" things he says. Frog announces why he went to the island in the river: he had woken to a shining sun, and felt so good that he had Toad for a friend, he had to be alone to think it all over. The book closes with a drawing of the two friends, sitting alone on the island, eating lunch together. They are seen from the back, arms around each other, gazing into the sky.

These books are not compendia of religious teachings. They are, however, moral teachings. What we have here is a very grown-up set of emotional issues made child-sized, in the recognition that children—even young children—may fret about mattering to others. The Golden Rule permeates these books, and we're meant to learn from the animals that espouse it, even as we enjoy their stories and the pictures of them playing and eating cookies together.

Centuries before Arnold Lobel wrote the *Frog and Toad* books, tales starring animals as moral messengers were hugely popular. Medieval bestiaries are one example, but the first and most famous of all were the tales collected by Aesop. What have come to be known as "Aesop's Fables" are tales about animals that, through nuanced use of animal characteristics, invite us to think about our actions in the world.

Of the man Aesop, questions abound and answers are few. He lived a life of enslavement in ancient Greece around 600 BCE; that much we do know. Was he of African origin, which would account for the many animals of African (rather than European) origin in his tales? Is it true that Socrates labored to put Aesop's fables into verse as he awaited his own execution? Probably the answer to each of these questions is yes, but we don't know for sure. We know Aesop's animals far better than we know the person behind the tales.

Many of the animals in this book grace the pages of the fables:

apes, monkeys, lions, elephants, and dogs are all there. After more than 2,000 years, the stories still engage us. Almost everyone knows at least one or two of the tales. "The Tortoise and the Hare" is the beloved story of the speedy, braggy, and supremely overconfident hare and the slow, deliberate, never-give-up tortoise, and guess who wins the race? This is the kind of wisdom we pass on to our children, not through abstract principles but through embodied animals, because animals bring to life the virtues that we want our families to live by.

In a notable parallel to the origin myths and animal teachings of Africa, Aesop's tales (at least as translated into English) hone in on specifics of animals' temperaments and abilities. In "The Dancing Monkeys," a prince trains some monkeys to dance, monkeys that are "naturally great mimics of men's actions." The animals are dressed in clothes and masks, and dance in a "spectacle" that meets with "great applause" from the audience. Then someone throws nuts in front of the dancing monkeys, who immediately discard their finery and surge toward the treats. They fight over the food, and no longer does the audience applaud, but instead erupts into laughter and ridicule. The lesson? Appearances may deceive.

In "The Crow and the Raven," the raven is considered a good omen by people who watch its flight to figure out the "good or evil course of future events." (Here is a rather realistic reference to raven augury!) This state of affairs makes the crow exceedingly jealous. The crow flies high into a tree and caws loudly. People look up, only to realize it's a crow (not the raven), and dismiss its cry. The lesson in this fable? Being other than your true self leads only to bad outcomes.

Aesop's fables instruct children on the proper virtues, but also invite discussion and reflection on the part of adults. The lessons may be clean and true, but they aren't simple, any more than Arnold Lobel's stories are simple: these tales tell of challenges, challenges that sentient creatures face together. By no means are these stories *only* for children. Animals as teachers, as able to speak to our deepest human needs for finding our moral compass in a complicated world, are found in the widest possible spectrum of chronological periods and

cultural spaces. Twenty-first-century versions of Aesop's tales sometimes pop up on my computer screen when I check my e-mail. Family
and friends delight in sending me online photo albums of animals, pictures initially compiled by some thoughtful stranger and then widely
circulated, or stories from the media that celebrate the warm lessons
we may learn from animals. These items invite us to think about our
own behaviors and choices in a too-often denatured and stressed-out
world.

Overwhelmed at work? Have a lie-down and relax, exhorts a dog
flopped over on its back or a heavy-lidded cat lying in a patch of sun.
Need a friend? Look beyond the obvious and the ordinary, and read
about the monkey and the pigeon, or the orangutan and the tiger,
oddly matched cross-species friends who saved each other from loneliness.[28] A staple of these online offerings is the idea that *we can learn
to love—our lives, each other—with and from animals.*[29]

When we step back and look at the big picture, we see that in
ancient texts and in online news stories, in lighthearted and serious
ways, animals usher people into reflective states that may spill over
into the religious. Sometimes, in a particular moment in a particular
place, there's an intensity that gathers force and changes the larger culture forever. Medieval France's obsession with bestiaries is one
example; the perspective of the nineteenth-century American
Transcendentalist writers is another.

The Transcendentalists' writings were an intense celebration of
nature's divinity. Indeed, the heart of Transcendentalism was the
belief that the divine could be found not in an external sky God but
within each of us, a divinity that could be nourished by connecting
with nature.

We know from his journals, kept at Walden Pond in Concord,
Massachusetts, that Henry David Thoreau saw "strange affinities" in
this universe. When he heard the "far off lowing" of a cow, it seemed
to him to "heave the firmament," and he suspected at first that it
was the voice of a passing minstrel. But it wasn't a human's voice, it
was a cow's, which is "but one articulation of nature."[30] It was

this unity, this mutuality, available to any person who walks out in the natural world, that the transcendentalists aspired to experience and understand.

"All are needed by each one; Nothing is fair or good alone." With these words, in the poem "Each and All," Ralph Waldo Emerson speaks to that sense of mutuality. For the Transcendentalists, the sacred inhered not just in an animal, but in an animal living in its right context: out in the forest, the desert, the mountains, surrounded by the sun and sky and other animals in the full beauty of where it was meant to dwell. Moved by a "sparrow's note from heaven," the poem's narrator takes the bird from its wild home and brings it indoors, only to discover that the pleasures of the bird's song die on the vine: "He sings the song, but it pleases not now / For I did not bring home the river and sky."[31]

A particularly beautiful example of mutuality, one that directly pairs animals and God, comes from Margaret Fuller. In "Meditations, Sunday, May 12, 1833," Fuller describes a scene of "clouds marshalling across the sky" and lighting the surrounding hills. On a breeze comes a "refreshing shower," perhaps of orchard blossoms, perhaps of rain borne in by the clouds. "The birds pour forth / In heightened melody the notes of praise / They had suspended while God's voice was speaking."[32] I imagine a kind of conversational turn-taking here: the birds tune in to God's shower, then duet by adding their song in their turn.

Growing from the Transcendental movement and ushering in the peak of nature-and-religion literature in nineteenth-century America was Walt Whitman. For Whitman, the Earth was as divine as heaven. Everywhere could be found souls, in animals and trees as well as people. So much resonates here for the anthropologist of the world's religions! Whitman's feelings echo across time and across cultures from North America, South America, Australia, Asia, and Africa.

Sometimes, as with his "A Noiseless, Patient Spider," there's a description of nature's beauty with an analogy made directly to humans. Just as the patient spider launches forth "filament, filament, filament, out of itself," the human poet-self, addressing his own Soul, describes the ceaseless "musing, venturing, and throwing" that goes on as the Soul awaits "the gossamer thread" to "catch somewhere."[33]

Yet Whitman's works are so rich, so suffused with longing and sexuality and observations about human emotions, it would be folly to think of them as "nature poems" unless we honor that term with the deepest and broadest and best evocation of the human condition. Whitman, perhaps especially because of his *Leaves of Grass,* came to be viewed as a poet-prophet, a religious figure even more than a brilliant man of letters.[34]

It's a staple of high-school English-class assignments to conjure in the mind's eye (and on the essay page) a meeting with a great man or woman of science, politics, literature or other arts, and imagine the conversation that would result. Inevitably I ponder the joys of discussing apes with Charles Darwin, but I long too for an hour's talk with Walt Whitman.

Primarily, I'd be noiseless and patient. I would listen to whatever thoughts Whitman would care to share.

Surely, though, I would steal a few moments to read aloud Craig Childs's memories of the feather-caching ravens in red-rock Utah country. Somehow I don't believe that Whitman would be incredulous or skeptical, or recommend that scientists confirm the observation by study of other raven populations. It's only a fancy on my part, of course, but I'd like to think that Whitman would listen and nod— and in the nod would be a tribute to "the unfailing perception of beauty and of a residence of the poetic in outdoor people."[35]

I do not believe that Thoreau's or Fuller's or Whitman's writing represents a uniquely elevated expression of humankind's seeing the beauty in nature. One can admire, one can love with heartfelt immersion, these great writers' prose and poetry, and yet resist the trap of equating it with a cognitive state higher than what "everyday" people like you and I may experience.

For one thing, Whitman believed that people who dwell in nature apprehend its beauty just fine all by themselves; they have no need for writers to translate the wonders of the natural world for them. He might say (and I would definitely say) that reflective practices by all peoples, traditional and industrialized, religious and nonreligious, in the present and the past, may be as deeply appreciative of nature as anything composed by schooled intellectuals.

Indeed, for the American writers I've mentioned here, the stirring spirituality of nature was not experienced primarily through the words on a page, as it may be for us today as we read inside our classrooms or suburban homes or city apartments. It was, instead, experienced in the outdoors life, or even just the evening walk or the hour spent listening to birdsong in the yard. Here again is continuity. From dogs that dream, to reindeer that choose to die for their human companions, to shamans' animal guides, animals ignite the spirits at large in the world or inside the human heart, in locally variable but equally intense ways.

7

Of Whales and Tortoises

*   *   *   *

All of interlocked life [is] talking at once . . .

—*The Echo Maker* by Richard Powers

OUR OBSESSION WITH ANIMALS is ancient. Just as our relationship with animals today goes far beyond eating them or using them for labor, so it was right from the beginning for *Homo sapiens*. Ever since we evolved into existence as a species, we have responded in multilayered, emotional, and sometimes spiritual ways to animals: this is the message of the book so far.

Now, I want to move more firmly into the present. What did all those millennia of evolving-with-animals prepare us for? That is, how are we transformed by our encounters with animals today? What are animals doing that so fascinates us even in this age of high-tech distractions, when we may retreat to our computer screens and generate whole worlds of fake animals to play with? Why do we still crave time with real-life animals? How has the modern science of animal behavior allowed us to better understand them?

In the next chapter, I'll consider these questions through the lives of the animals I know best, the monkeys and apes whose primate selves reflect our own lives back to us, with some intriguing distortions.

But it's fun—and instructive—to consider first some animals more distantly related to us.

At first glance, whales are so "other" compared to humans that they might seem to be impenetrable to our comprehension. Navigating through vertical depths with grace and speed, whales inhabit a watery realm we can only dip into briefly, unless we augment our own skin with special wetsuits and oxygen tanks. Huge creatures, legless and handless, with no easily recognizable gestures or facial expressions, whales represent such a departure from a more familiar four-legged mammalian body plan that by all rights they should be opaque to us primates. Surprisingly often, this isn't the case at all; sometimes whales let us peer into a heart and soul ignited by being with animals.

In the book's opening pages, I described what happened when a humpback whale became trapped in strong crab-pot ropes in the waters off San Francisco. After divers concluded the arduous process of freeing her, she behaved in a fascinating way. She approached and bumped up against each of her rescuers, one by one. Diver James Moskito expressed how deeply moved he felt ("it was an amazing, unbelievable experience") at this action by the whale, one he judged to be an expression of gratitude.[1]

Moskito's experience surely counts as mutual relating with an animal! That much is obvious from his own words. That some people suspected a "fish story" is suggested by the appearance of this whale-human encounter at snopes.com. This website assesses the stories that we hear all the time, never quite knowing if they're based on wild rumor or plain fact. These tales are sometimes sobering, sometimes amusing. Two examples judged to be false: "Hotel room keycards are routinely encoded with personal information which can be easily harvested by thieves," and "The average person swallows eight spiders per year." The whale story is followed by a terse "Status: true."

But of course, this judgment begs all the most intriguing questions. "True" means that the media's reporting of the rescue, including the whale's behavior, is deemed to be accurate. But *was* the whale grateful? Was she reaching out with the hope of expressing something to her human rescuers? According to the May 2006 issue of *Reader's Digest*, at least one scientist tried to explain the whale's behavior in strictly practical terms: perhaps the great mammal craved exercise after the confinement of the nets, and the divers just happened to be there as she moved through the water. The divers, the people in the water next to the whale, believe that much more was going on. And they aren't the only ones.

The writer Lynne Cox describes a longer "whale event" using terms that are unmistakably spiritual.[2] One early March, also in California, on a morning when the "sea and sky were inky black," Cox went swimming off Seal Beach. Seventeen years old and an experienced swimmer, she felt something unprecedented happening in the water around her: "The water began shaking harder than before and I was being churned up and down as if I was swimming through a giant washing machine. The water shifted, and I was riding on the top of a massive bubble. It was moving directly up from below, putting out a high-energy vibration. I felt like there was a spaceship moving right below me."[3]

Only when a friend on shore called out to her did Cox learn that a baby gray whale had been tracking her for a mile. Later still, Cox figured out that the unusual vibrations had come from his mother, and that the two whales must have become separated somehow. At that time of year, gray whales migrate from Mexico up to the Bering and Chukchi seas, and this one was too young to be on his own.

Over the next few hours, Cox tried to figure out how to reunite the baby, whom she named Grayson, with his mother. She eventually succeeded, but it took time, and while she was in the water she fell under Grayson's spell: "I couldn't take my eyes off him. He swam within ten yards of me."[4] Time and again, Cox's emotions spilled over, perhaps most acutely when it seemed as if Grayson and his mother would never be reunited.

The emotion that Cox felt burns through her words about Grayson: "He looked so small in the enormous sea and I wanted to protect him somehow. Maybe you communicate with your heart. That is what connects you to every living thing on earth. Use your heart. It is love that surpasses all borders and barriers. It is as constant and endless as the sea. Speak to him with your heart and he will hear you."[5]

My response to this passage is split. It's right on the money in terms of modern-day being-with-animals. We humans, after a long evolutionary journey as and with animals, are poised to open up to the creatures around us—even when we live in urban societies. Yet the scientist in me reads four key words in that passage with acute skepticism: "he will hear you." It's a lovely thought, that the whale could hear Cox's goodwill. Taken literally, it is not realistic. Barriers to communication are formidable even within our own species. How often do we sit in a room with a cherished but troubled member of our family, and will her to hear what is in our heart? Often, despite our profound wishing, it doesn't work that way. Communication across species surely involves extra layers of challenge. In short, I cannot subscribe to the view that we humans can beam our desires and emotions so that whales (or apes or elephants) may hear them.

Here's what I do believe: that the desire and emotion we have in our heart invests our every movement and gesture and expression and voice tone and word with a meaning that can be *felt* by another creature.

Cox writes, following the passage I quoted, that Grayson looked at her, and seemed to await her following him. I think that Cox's actions communicated something to Grayson, a creature whose species had evolved to be socially attuned. Even at a young age and even though his partner was a human, Grayson was able to pick up on Cox's cues. I do not mean to suggest that Cox conveyed to Grayson a concept or idea, but rather that she embodied, by the way she moved and held herself in the water, a message on the order of "no danger here" or "I am safe to be with."

Cox and Grayson then swam toward the area where Cox figured Grayson's mother might have swum. After some false leads and plunging hopes for Cox, and fascinating behaviors by Grayson, the mother whale did appear. She gave off a series of vocal calls, and Grayson replied. Mom and baby were reunited.

The story doesn't end there. The mother "was massive and it was amazing . . . [s]he knew how close she could get without swimming down something as small as me. . . . She was right beside me. For a moment, I touched her cheek. . . . These was a glimmer of light in her big brown eye. I felt a connection between us, just as I had with Grayson. She looked at me. I looked at her. We held each other's gaze. It seemed like she was saying thank you."[6]

Just like the humpback whale in the waters off San Francisco, this gray whale participated actively in an encounter with a human—as, indeed, did Grayson himself. The mutual relating pierced all species boundaries, apparently, for all three beings involved. Cox writes that she will always remember that morning in the ocean. "Emotional" is simply inadequate for describing Cox's sense that, at that place on that morning, she wasn't alone in the huge ocean, just as she isn't alone in the infinite universe. "Spiritual" takes in all that emotion and recognizes something more.

Soon after I read Cox's account, I came across a wonderful passage in Margaret Drabble's novel *The Sea Lady*. Set in Britain and infused with luminous language on the theme of marine biology, at its end the book offers a character called Humphrey who muses about the nature of forgiveness. As I read, in my head I replaced Drabble's word "forgiveness" with "compassion": "[I]t comes to [Humphrey] that forgiveness need not be maintained in time. It may come in an instant, like grace. It need not endure. One may be redeemed in an instant. Repentance needs only an instant, a measurement too small to show on the clock face."[7]

Repentance, redemption, grace. Maybe compassion, too, can come to a person in an instant, like grace. That it could is a hopeful thought. How much more hopeful to imagine, and to intend, that compassion can endure.

There's no certainty that being with animals will lead to

compassion, either in one's thinking or in one's practice. *But sometimes it will; the potential not only exists, but exists uniquely when animals are involved.* Indeed, change may come even to those not directly involved in an amazing encounter. Why has the story of the divers' encounter with the whale off San Francisco gotten a great deal of attention in the press? Why has Lynne Cox's book about Grayson been praised by Jane Goodall, the renowned chimpanzee expert? Goodall commented, "Everyone who reads *Grayson* will be enchanted and profoundly moved. *Grayson* is a powerful voice for conservation."

Indeed, projects aimed at saving whales—entire populations or single individuals—have tapped a chord of emotion in animal lovers. *Free Willy*, a hit movie of 1993, told the story of a young boy and a whale. The latter was played by Keiko, a killer whale owned by an amusement park in Mexico City. When Keiko became a star, thousands of people, young and old, mobilized to free him from captivity. After a multiyear, multimillion-dollar effort to prepare Keiko for survival in the open ocean, he joined wild whales in 2002. To the joy of many, Keiko swam free for nearly 1,000 miles on a journey from Iceland to Norway. Sadly, Keiko died the next year, most likely of acute pneumonia, although old age may also have been a factor.

Let us consider one more example of this type of emotional connection. Once again, individuals of two different species form a close bond. This time, however, neither species is human.

OWEN AND MZEE

The friendship between Owen the baby hippopotamus and Mzee the giant tortoise has captured the imagination of people around the globe. Two books recount the story. Though intended for children, *Owen & Mzee: The True Story of a Remarkable Friendship*, and its sequel, *Owen & Mzee: The Language of Friendship*, are compelling for adults as well.[8]

Few of us can forget the intense destruction and loss of life caused by the "Christmas tsunami" of 2004. The facts as recounted in a science journal convey a terrible tale:

> At 0:59 GMT on 26 December 2004, the India tectonic plate slid underneath its neighbour, raising it by about 10 metres. The earthquake caused a powerful tsunami that devastated coastlines around the Indian Ocean. Entire towns were wiped from the map, almost 300,000 people were killed, and millions have been left in urgent need of food and shelter.[9]

This disaster disrupted the lives of countless nonhuman creatures as well, and here is where Owen comes in. Earlier that same month, a group of about twenty hippos living along Kenya's Indian Ocean shore were washed downriver by heavy rains. They ended up right near a village. Hippos are notoriously dangerous. Unpredictable by nature, they can become quite aggressive to humans, so this development was not a welcome one for people living in the area. The villagers' attempts to chase the hippos back upriver ended in failure. Among the hippo group, they noticed, was a baby male, young enough to cleave to his mother's side.

And then the tsunami hit. Fortunately the villagers who were out on the water at the time, fishing, were rescued. Soon people noticed that in the turbulence, the small hippo had become separated from his mother, and indeed from any other hippos. He was "stranded on a sandy coral reef among the sea grass. Tired and frightened, he was unable to reach the shore on his own."[10]

After a lengthy and dangerous struggle, during which the 600-pound youngster was quite angry and upset, rescuers managed to transport the newly named Owen to Haller Park, an animal sanctuary near Mombasa.

I've never been to Haller Park. From my days in Kenya, I can report that Mombasa is an inviting city, a city that retains hints of its former life as a vital way station on the ancient spice-trade route. Today, syllables of Swahili linger in the air as foreign resort guests mix with locals on bright white beaches that border warm sea waters. Fittingly, at Haller Park, hippo Owen was invited to enjoy the water, too. He was brought to an area with a pond and a mud wallow, where he joined resident bushbucks and vervet monkeys. And Mzee, the giant tortoise.

Now, Mzee's situation was much different from Owen's—for one thing, he was about 130 years old when Owen arrived! Imagine, an animal making his slow and steady tortoise way in the world since well before the British government declared Kenya and Uganda to be a protectorate (1894). And imagine, in the more familiar framework of American history, an animal born shortly after the Confederate surrender at my home state's Appomattox Courthouse (1865), the event that brought the bloody battles of the Civil War to a close.

In recent times, Mzee had been pretty much a loner, though he did seem to like one caretaker at the park. How Mzee's slow-paced life must have tilted when suddenly Owen rushed off the rescue truck, directly toward him! "Owen crouched behind Mzee, the way baby hippos often hide behind their mothers for protection."[11] Mzee's immediate reaction was all too obvious: he hissed and moved away from Owen.

At some point during the night, though, Owen huddled against Mzee, and Mzee *didn't* move off. Owen began to eat next to Mzee. There's no mistaking the fact that Owen was the instigator in this relationship: when Mzee would crawl away, Owen would follow. Once in a while, though, even during these early days, Mzee would follow Owen instead.

Time passed. Owen grew, and so did the bond he shared with Mzee. The bigger of the two and certainly the faster in terms of possible cruising speed, Owen still took his cues from Mzee: when Mzee ate, Owen ate. When the two roamed the enclosure, Owen matched his pace to Mzee's.

As someone who has studied social communication in primates for over two decades, I find Owen and Mzee's unique system of communicating to be remarkable. They rumbled back and forth to each other in a way typical for neither hippos nor giant tortoises. And their bond is literally embodied in the tactile communication system they created together: "When Mzee wants Owen to walk with him, he will gently nip Owen's tail with his sharp beak. When Owen wants Mzee to move, he will nudge Mzee's feet. To direct Mzee to the right, he will nudge Mzee's back right foot. To direct him to the left, he will nudge Mzee's back left foot. If Mzee doesn't respond right taway, Owen may squeeze Mzee's foot between his teeth until he starts to move. But neither ever hurts the other."[12]

Once again, we see transformation across species boundaries! And the transformation extends even beyond a hippo and a tortoise, because when people learn about Owen and Mzee, they often get caught up in the story in the same way people were caught up in Keiko's. Perhaps this response to a cross-species friendship may be explained by the tsunami. It's a welcome relief to learn that something good—even something small, in the scheme of things—emerged from an event so life-destroying.

I think there's more to it, though. This true story has caught fire among adults and children alike. Aesop couldn't have invented a better teaching tale: persist through adversity (the tsunami that separated Owen from his mother), stay open to life's possibilities (in meeting another creature not so much like you), and good things will blossom (such as the close friendship between Owen and Mzee, leading to a happier life for both animals).

Consider one reader's response to an Owen-and-Mzee book, posted on Amazon.com: "As my 7 year old and I read this beautiful story together, we had to take turns finishing each other's sentences because of overwhelming emotion. She was so inspired by their friendship, she immediately requested to go to any Tsunami-devastated country to rescue and take care of injured and orphaned animals! Our fellow creatures continually amaze me in their capacity to inspire and nurture our souls."[13]

Explicitly emotional, and infused with compassion, this comment reveals one little girl's wish to make a difference for other creatures, a wish brought alive by a true story. Owen and Mzee affirm the very thing for which we, in our human-made world of strife, most yearn: the knowledge, even more the belief, that we *can* care for others different from ourselves, and that together, through mutual will, we *can* overcome barriers to shared communication.

Of course, it's not every day that we encounter whales in the ocean, or come across a story of two animals bonding in an exceptionally moving way after a disaster. But animals grace our lives every day—in my case, in Africa, Atlanta, and points in between.

8

Articulate Apes and Emotional Elephants

• • • •

Once, after a few days of heavy rain, we stumbled upon a
plethora of newly emerged mushrooms—a baboon
delicacy that normally evokes competition. This day,
however, there were enough mushrooms for everyone. To
my amazement, before anyone dug in, they all paused to
join in a troop-wide chorus of food-grunts, their bodies
literally shaking with excitement. In that moment, I
realized that collective rejoicing in celebration of
sustenance must have begun long ago.

—Barbara Smuts, "Encounters with Animal Minds"[1]

SOME YEARS AGO, IMMERSED in writing anthropology articles
about ape communication and the origins of language, I flew
to Atlanta, full of anticipation. I drove from the airport to a rather
remote fifty-five-acre compound, the home of Georgia State
University's Language Research Center. There I caught a glimpse of
Kanzi the bonobo, a large and serious-looking ape famous among
those who study language.[2] I approached his enclosure with mount-
ing excitement, and greeted him with a "Hello." The first words Kanzi
uttered back to me were, "Visitor, chase."

Naturally, Kanzi did not verbalize these words, because apes are not capable of speech. Rather, he pointed to an image on his portable symbol board, and scientist Sue Savage-Rumbaugh translated into English for me. Kanzi dubbed me "Visitor," and then asked me to join in his favorite game. And chase we did; Kanzi ran along the fence inside his enclosure, while I ran just outside it.

Knowing that Kanzi comprehends a good deal of spoken English, next I remarked to him, "Kanzi, I brought a gift for you." I indicated to him the outlines of a small object bulging from the side pocket of my blue jeans. Looking intently at this shape, Kanzi used his symbol board to query, "Egg?" Kanzi was wrong in guessing that I had brought him an egg, but his mistake was both intriguing and impressive. The object in my pocket was a small ball, close in shape to an egg.

I showed up at the Center having done my homework. From the academic literature, I knew that the images on Kanzi's board were not representations of the objects or concepts in Kanzi's world, but instead were symbols. Nothing about the four-letters-strung-together of our English word *play* represents the actions of play (any more than the French, Dutch, Arabic, or Chinese words for *play* do). By the same token, the image for *play* on Kanzi's board is wholly arbitrary: a rectangle, outlined in green and divided by lines into three equal parts.

Most fascinating to me was the fact that Kanzi had not been trained in the use of these symbols. Years before, young chimpanzees and gorillas at other research facilities had been drilled in aspects of the American Sign Language used by the deaf. Trainers had molded the hands of these apes into the appropriate symbol shapes, and had given their subjects rewards of food when they imitated new signs correctly. Some of these apes did use ASL creatively in conversation.

Kanzi's history was different. As an infant, he had scampered around while his mother, Matata, was being given lessons by researchers at the symbol board. By all accounts, Kanzi had paid no attention, preferring to play rather than attend to the serious business at hand.

When Kanzi was two, Matata was temporarily transferred else-where for breeding purposes. On the very first day of this separation, Kanzi used the symbol board to produce for his human companions 120 short utterances. Over the next years, Kanzi's accomplishments rocked the ape-watching world. Rigorous experiments attested to his growing proficiency with both the symbol board and comprehension of spoken English. The results flabbergasted scholars across academic disciplines, and even won over some of the harshest critics of ape-language work. Kanzi proved that immersion from a young age in a language-rich environment can have amazing consequences for an ape. To a surprising extent, Kanzi's upbringing reflected the processes of language-learning that any child experiences.

Marvelously enlightening for an anthropologist, my encounters with Kanzi were unforgettable. Far more powerful than any intellectual rush was the emotional one. There I was, no longer reading about what Kanzi could do, but joining with him in communication!

To say that being with Kanzi was a thrill doesn't go far enough, however. It doesn't capture the sense of wonder and connection that flowered within me when I was with Kanzi. During those times, I knew that I was in the presence of a keenly conscious being, an individual of another species with thoughts and feelings all his own. I was relating with a creature who was at once *other*, and yet quite like myself. It was an expansive experience, an intensely satisfying one, and one that helped me *feel* the beauty of the idea that all living species on earth are connected in a web of relating, a web that extends into the realms of intelligence and emotion.

The scientific literature tends to anoint Kanzi with nearly unique status in the animal world. This isn't because other apes lack the capacity to do what Kanzi can do; indeed, his half sister, Panbanisha, may exceed him as a linguistic superstar. It's because only a very few apes have ever been challenged in this way to show off what they know. The work is fascinating and adds to our understanding of the linkage between rearing system and cognitive expression in primates. Yet there's something to be said, too, for meeting apes on their own terms: no human-devised communication systems, just observation of what apes do on their own.

BONOBO WORLD

My own research at the Center involved Kanzi only peripherally. Instead, I observed his mother and his three younger sisters Tamuli, Neema, and Elikya, none of whom were language-competent. I was fascinated with their subtle attention to each other's body language and gestures. Occasionally Kanzi or Panbanisha would visit their mom and siblings, and that was always fun, but I focused mostly on Elikya: my student Erin Selner was lucky enough to film her birth and we watched her develop right from day one.

One thing that caught my attention was how Elikya responded to the caring watchfulness of her older sister Tamuli. Owing to a congenital heart condition, Tamuli had never been mated, and had no infants of her own. She was, though, a careful and attentive older sister. Mother Matata and sister Tamuli each taught Elikya how to walk. The two older bonobos walked backwards, quite deliberately, just in front of Elikya, urging her on with their movements as she toddled along. Mom and sister also guided Elikya in learning various bonobo behaviors and family customs.[3] Skirmishes occasionally broke out among the four apes, but I hadn't fully realized the degree of their harmony until the day Panbanisha first came for a visit while I was there.

As I have mentioned, Panbanisha is Kanzi's half sister and was at that time an emerging linguistic star in her own right. Yet when she visited this group, she became just another of Matata's daughters. She did not seem to harbor much affection for her littlest sister, Elikya. I witnessed no fond embraces between the two (though it must be noted that I did not keep a round-the-clock record). Twice, as Panbanisha strolled past Elikya, out flew her leg for a swift kick to Elikya's head. The first time I saw this, I wondered fleetingly if

Panbanisha had somehow stumbled or lost her balance. But no, it seemed quite a determined and deliberate kick, and then it happened again.

Had the apes spent more hours together, familiarity might have bred better older-sisterly behavior! As it was, the contrast between Tamuli's actions and Panbanisha's was instructive to me. Here were two apes, equally related to Elikya genetically, yet expressing themselves in highly different ways around her.

Gradually, as I observed Kanzi and Panbanisha conversing in symbols with their caretakers, and studied the ways that Elikya's sisters interacted with her, a stunning principle was becoming clear to me: *There is no innate bonobo nature. It's not only that apes differ from one another in their personalities; it's that no individual's trajectory is fixed at birth. Just like human children, primate infants are deeply influenced by their environments. Just as there is no fixed and inviolable human nature, there is no fixed and inviolable ape nature.*

The malleability of ape "nature" is evident in Kanzi's accomplishments: just as his Swahili name promises, he represents "buried treasure" for the scientific community. Reared in a highly enriched environment, he has absorbed more from the actions of others around him than anyone had dreamed possible for an ape. But the malleability principle is bigger than just Kanzi, or Kanzi and Panbanisha—bigger than just bonobos.

THE PLEASURE RUMBLE

To walk onto the grounds of the Smithsonian's National Zoological Park in the early morning is to experience an audible feast. With a bit of luck and a favorable wind direction, you can hear the songs of Asian apes, the calls of birds of prey, maybe even the quiet munch of camels at work eating their breakfast. Despite the pleasant symphony, I walked quickly down the path and through the sound waves; I had primates to see.

It was now two years after the Elikya Project in Atlanta had finished up. Commuting to Atlanta from Williamsburg, Virginia, wasn't
really a viable long-term strategy for my students and me, and when I
was invited to study the western lowland gorillas closer to home at the
National Zoo in Washington, D.C., I felt very fortunate. During my
visits, I greeted each gorilla, one by one. All of these large-bodied and
magnificent African apes can distinguish a menacing voice from a
friendly one, and a relaxed body posture from a tense one. This fact
will surprise no one who spends her day with a dog or a cat. From so
many years in close relationship with humans, each ape also comprehends some English words. In Atlanta I'd once met a male bonobo
who had recently arrived there from Japan. It was fun to watch him
respond to Japanese words! As we have seen with Kanzi, apes may
truly surpass other mammals in this regard, and apes can understand
a social companion's perspective on a given situation in a way that
most mammals cannot.

Through a combination of speech (by me) and gestures, body pos-
tures, and vocalizations (by all of us), the gorillas and I interacted.
Many of these interactions feel quite different to me than even my
most in-sync moments with my favorite cat. With the gorillas sometimes I am directly engaged by brief eye contact and gesture, and
sometimes I am charged by a body hurtling itself stiff-armed and
hair-erected against the cage bars in a fearsome display of power.
Occasionally I am ignored. In all these cases, what comes across is
that I'm engaging with another being who approaches the world in
thoughtful—though sometimes quite impulsive—ways, and who
experiences that world through hearing, taste, touch, smell, and, most
of all, vision, eerily like my own.

For years, at each of my intermittent visits to film and study the
Great Ape House gorillas, I greeted Kuja first, the group's weighty sil-
verback male. This etiquette was only appropriate. In captivity just as
in Africa, gorilla groups are organized along fairly strict dominance
lines; an older male, bearing age's badge of silver hair, takes charge.
(Sometimes two males share the leadership role.) Kuja, an imposing

ape with his powerfully muscled body and large canine teeth, was the undisputed power center of the group I chose to observe.

During the years I filmed the gorillas' behavior, five other apes shared Kuja's living space. I enjoyed watching his companions—even his longtime mate Mandara and his sons Kwame and Kojo—move rapidly around the enclosure while managing to orient their bodies and eyes toward Kuja. When Kuja assumed a tense position, the others instantly became alert and tensed themselves; when the big guy relaxed, they too relaxed. The gorillas had perfected the embodied art behind the principle "don't turn your back on the boss!"

I looked forward to greeting Kuja. Almost always, I supplemented my verbal "hello" with a version of a gorilla pleasure rumble. This vocalization emerges from deep within the gorilla and indicates . . . well, pleasure! Enjoyment of food, perhaps, or pleasant acknowledgment of another's company. Almost always, Kuja would rumble back to me—and was forgiving of my human accent.

Kuja rumbled frequently with his caretakers, of course, who spent far more time with him and knew him far better than I. But he wouldn't rumble with just anyone. I felt proud to be so chosen by him and felt a lightness in my chest, a happiness suffusing my body, when he rumbled with me.

Without a doubt, Kuja recognized me as an individual. When a large crowd gathers around the large glass window at the gorillas' indoor enclosure, or the rail at the outdoor yard, the apes routinely pick out with their eyes a member of the zoo staff, or me, if I am present. Mandara, the adult female mated with Kuja, often looked at me and opened her mouth wide. In this way she requested food, a request she didn't bother to make with garden-variety zoo visitors.

Mandara was the gorilla who made me think the most about apes' ability to distinguish people by gender. One day, the then-dean of Arts and Sciences at William and Mary joined me at the Great Ape House, along with his wife and a mutual friend. Immediately, Mandara began to direct behaviors toward the dean, the lone male among us. Ignoring all three females, Mandara pursed her lips and strutted in front of him.

I began to feel dry-mouthed and uncomfortably warm. I knew full well that she was exhibiting not only curiosity but sexual interest—in my dean! Here I was, putting the person who was more or less my boss into a situation of being flirted with by an ape! Next, Mandara pushed a small piece of bamboo through the cage mesh, again clearly aiming at only one person among us. To my great relief, the dean found all this amusing and even touching. He kept the bamboo as a small memento.

How could Mandara suss out the lone male in our little group? The dean is a tall man, but I am a tall woman. He has a beard and a deeper voice than mine or the other women's. Whatever signals Mandara employed to determine gender, her choice that day of pre-ferred human companion was clear.

When gorillas interact selectively with certain individuals whom they have come to know, it's not just owing to the tip-off of a zoo uni-form or because a person is working in a particular location on a cer-tain day. During an early-career summer fellowship at the National Zoo, I studied orangutans, but the gorillas were nearby and I got to know them, too. One gorilla male took a serious disliking to me. I have no idea why, any more than I understand why Mandara liked the look of my college dean. This male charged and threw objects at me every day. The "objects," unfortunately for me, included products of his own body (to put it delicately). Equally unfortunately, gorillas can be quite accurate with their throwing arms.

One day a zookeeper urged me to take revenge and turn a hose on the cantankerous gorilla. The idea was to establish my dominance over him, or at least get some respect. (The present-day zoo staff would not suggest such a thing, by the way, and my present-day self would not agree to hose any ape, grumpy or otherwise.) In any case, the hosing made no difference in my relationship with this gorilla.

One weekend, I brought my parents and two friends to visit the Ape House. It was my day off, and upon arrival we went only to the public area, not to the location where I collected data every day. As we stood in a thick crowd, this same male shifted from a normal relaxed posture to that of an ape possessed. Immediately he stared at and charged toward me, in a manner so directed that zoo visitors standing around me took notice, and perhaps wondered what I had ever done to this poor ape! I was being responded to as a specific person who shared a history with this ornery gorilla. Any primatologist or zookeeper will tell you the same, that he or she is known to the apes as an individual.

But just as in Atlanta I had focused on an infant, so in Washington did I want to do the same. My first gorilla "star" was Kwame. Both Elikya and Kwame used their bodies, limbs, heads, and eyes to coordinate their actions skillfully with their mother and other groupmates. Each infant's behaviors were rooted in their intimate relationship with a caring and competent mother: Matata (bonobo) and Mandara (gorilla) both were experienced "supermoms" of the captive ape world. From this secure base, the infants felt free to explore their world. Also, Elikya and Kwame both had older siblings to act as tutors of a sort. I knew, like any student of primatology, that first-time ape mothers, in their inexperience, often fumbled and made mistakes. Not all ape infants are as confident and communicative as these two babies.[4] Mothers are not all "supermoms."

For any infant, a unique web of factors related to inborn temperament, behavioral events within the family, and nuances of the lived environment makes up the matrix within which development occurs. Change any single variable, and the infant's behavior is affected, just as Kanzi would have turned out to be a different bonobo had he not been reared at Matata's knee while she was undergoing linguistic instruction, or if he had not been encouraged by his human caretakers to communicate through symbols.[5]

If living conditions can alter the developmental pathway for apes as much as I was growing to suspect, humans and apes were more alike than I had ever before realized.

EXPERIMENTING WITH EMPATHY

The sense of deep emotional capacities in apes that I was beginning to form through my own work matched up with news from the primate literature. Animal behavior scientists are increasingly intrigued by the question of what it *means* that highly social animals like apes can tune in to each other's emotional states and "read" each other's subtle cues and gestures as closely as I was documenting on film. Could apes, for instance, be said to experience and express empathy? Could they understand that a social companion was in need, or pain, or distress, and if so would they act to help?

A group of scientists in Germany has shown that chimpanzees help each other in the absence of any genetic relationship, and with no expectation of future reward. This amounts to true chimpanzee altruism, and relates closely to the question of ape empathy.

Three kinds of experiments were carried out. Two were designed to test whether chimpanzees aided humans in the solving of a task. Even though the chimpanzees were paired with people they knew very little, people who had never fed or rewarded them, the apes did choose to help. The third experiment I will describe in more depth; its goal was to find out whether chimpanzees would help other chimpanzees. Food was put into a room; chimpanzee number one observed as chimpanzee number two attempted to enter that room. Chimpanzee number two could not enter, however, because a chain blocked the door. This chain had been set up so that it could be released only by the observer chimpanzee, chimpanzee number one. Only if number one chose to release the chain would his partner, number two, get the food. And here's the rub—if that happens, number two gets *all* the food! The helper chimpanzee gets no food, indeed no reward of any kind, for aiding his partner.

The results were clear. Observer chimpanzees did choose to release the chain, and their partners enjoyed a snack. Importantly, the scientists note, the chimpanzees were "genetically unrelated group members, and the roles of recipient and subject were never reversed within a pair to exclude the possibility of simple short-term reciprocation in the same situation."[6] Further, the observer chimpanzees were more likely to help if their partner actively tried to get into the food-laden room (as opposed to just sitting there wishing for the food, apparently), and once the experiment was over, they didn't try to beg for part of the food. The conclusion is watertight, as far as I can see: the observer chimpanzees behaved altruistically, just because they wanted to help. In my book (no pun intended), that points towards empathy and altruism.

Empathy is the ability to walk a mile in someone else's shoes. With empathy, one chimpanzee (or person) understands something about another creature's mental or emotional state. It is, as the psychologist Paul Ekman puts it, a kind of emotion recognition. But the chimpanzee experiment I have described points towards more than only empathy. After all, chimpanzee number one, the helper, didn't simply *recognize* that chimpanzee number two might enjoy some good food. This chimpanzee went further, to bring about altruistically a happy solution for a social partner.

Often, skeptics "explain away" apparent altruism that occurs under natural conditions. These skeptics might point out that, in the wild, the helper ape is genetically related to the aided ape, or that the helper is likely to get some kind of short-term reward for his behavior. Yet consider an anecdote related by the primatologist Frans de Waal about the chimpanzee Peony. An older female living in an enclosure near Atlanta (the ape capital of the United States, clearly!), Peony suffered from arthritis. At times, walking and climbing were difficult for her. One day, "Peony is huffing and puffing to get up into the climbing frame in which several chimpanzees have gathered for a grooming session. An unrelated young female moves behind her, places both hands on her ample behind, and pushes her up with quite a bit of effort, until Peony joins the rest."[7]

To me, the chimpanzee helper's behavior seems likely to be altruism motivated by empathy. Is her action any less convincingly interpreted as altruism because she frequently interacts with Peony, and might in the future get her favor paid back by Peony? What if she had been Peony's relative? Would that mean the favor was entirely due to shared genes? These questions are complicated ones made even more complicated by the fact that scientific definitions (of altruism, for example) do not always track casual ones. I won't delve into the ins and outs of these matters, but instead recommend two books that illuminate them.

Emotional Awareness is based on an extended conversation between Paul Ekman and the Dalai Lama. Ekman distinguishes among emotion recognition (a process I described above), emotional resonance (when one *feels*, rather than only knows, another's state of being), and compassion (when one is compelled to act on emotional resonance in order to help the other). For the Dalai Lama, compassion is a matter of training; with self-discipline, a person who feels compassion has no choice but to act and relieve the suffering of another creature.[8]

In *Wild Justice*, the animal-behavior expert Marc Bekoff and the philosopher Jessica Pierce show that a whole host of behaviors related to empathy, altruism, and compassion exist in species as diverse as rats, dogs, and elephants. In fact, Bekoff and Pierce argue that sentient animals have morality—not some incipient proto-morality, but morality, period. Varieties of morality (and its building blocks) do exist, and their expression depends on a species' brain power, living arrangement, and natural environment. We humans have the most elaborated form, but Bekoff and Pierce argue convincingly that a scientific search for morality needn't concentrate entirely on the primates.[9]

It's clear from the literature as a whole that empathy may be thoughtful (cognitive) in nature, or it may be programmed in.[10] None of this alters the fact that chimpanzees *feel for* and *act for* companions who are struggling in some situation, at some cost to themselves. They want to help, and they think about how to help. Their behavior supports a conclusion of Marc Bekoff's: "[A]nimal emotions are not restricted to 'instinctual responses,' but entail what seems to be a good deal of conscious thought."[11]

Indeed, apes show a remarkable ability to share another's perspective—to grasp cognitively that another creature has his own way of viewing the world, and his own set of emotions. (This ability, called "theory of mind" in scientific parlance, takes years to develop in human children.)[12] One of the most intriguing new avenues to open up in the study of ape emotion and perspective-taking involves the response to death. Both wild and captive apes respond emotionally to the body of a dead companion, in ways different from how they react to an injured one. The response may involve empathy of a third party for those who have experienced the most loss, as happened in a forest in Côte d'Ivoire, in West Africa. When a female chimpanzee was killed by a leopard, the alpha male of her community kept all infants except the female's little brother away from her corpse. The brother sat with and gently touched the body of his sister.[13] Surely the alpha male acted that way because he understood something about the emotions that little chimpanzee was likely to feel.

In zoos across the United States, gorillas are now encouraged to spend time at the body of a newly deceased companion. When Babs the gorilla died at the Brookfield Zoo in Chicago, her nine-year-old daughter, Bana, was the first to approach the body. She held her mother's hand, stroked her mother's stomach, then lay down next to her mother and put her head on Babs's outstretched arm. Other gorillas came by, too, to touch the body gently. In her last few days of life, Babs had been quite ill, and the other gorillas had stayed unusually close to her. Babs's companions thus witnessed up close the decline in her condition. Finally, zoo staff stepped in and euthanized Babs, in a separate room, out of the gorillas' sight. I cannot help wondering what thoughts and feelings were experienced by the gorillas when they first saw Babs's body.

The gorillas stayed near Babs for a half hour, then began to file out. Last to leave was Bana. "She would get up," said one zookeeper, "move a few steps, stop, and turn back to stare at Babs. She started and stopped several times before she finally joined the others."[14]

At this point some of you may cry "Anthropomorphism!"

Anthropomorphism is the attribution of human qualities to animals—for example, assuming a gorilla is grief-stricken because her behavior reminds us of our own at a recent family funeral, or deciding a bonobo is jealous because her behavior mirrors the way your sister acts when she covets your new boyfriend. Anthropomorphism causes hot debate in the scientific world: Is it a damaging practice that romanticizes the animal world, or is it a technique that lends extra insight into the interpretation of animal behavior? I tend to agree with Bekoff's balanced view that "inappropriate anthropomorphism is always a danger," and yet "far from obscuring the 'true' nature of animals, [it] may actually reflect a very accurate way of knowing."[15]

On the subject of death responses in apes, Bekoff writes that "gorillas are known to hold wakes for dead friends."[16] Is this going too far? *Something* is going on when apes gather around the body of a dead companion, and it surely seems to involve grief. Could it involve more? Could it encompass an awareness of death, even awareness of the passing of a soul out of this life? No one can answer those questions. Far more incidents need to be observed before we can affix the label "wake" to them. We may never be able to make secure claims about the meaning of these behaviors to the animals themselves.

I am agnostic, too, about Jane Goodall's suggestion that wild chimpanzees express a kind of spiritual awe. Almost fifty years ago, when Goodall had been observing chimpanzees in Tanzania for only four months, she saw a spectacular rain display involving seven adult males: "I remember the sky over the slope opposite got darker and darker ... Then the rain arrived, the thunder crashed and rumbled from the purple-black clouds. Suddenly, opposite me, the chimpanzees climbed into low trees at the top of the very narrow valley. And then, one by one, the big males charged down the grassy, tree-studded slope, dragging branches in the drenching rain. One or two stood upright, holding onto low branches, swinging rhythmically from foot to foot. After his display, each one walked back up the slope, then started his display, his dancing, all over again. It was amazing. . . . What I saw was an expression of what I think is a spiritual reality."[17]

Goodall was to observe this dance again many times, both in the rain and, in similar form, at the waterfall near the chimpanzee camp. No way am I foolish enough to contest Jane Goodall's views on chimpanzees! When she asks, "Is it not possible that the chimpanzees are responding to some feeling like awe?"[18] I can only agree, *Yes, it is possible.* I wonder, though, if the awe response emerges as much in us as in the chimpanzees when we learn that our closest living relatives connect in an emotional way with their natural world. Is that connection *necessarily* a spiritual one? Why should it be, any more than a human being's is necessarily spiritual? Still, what a transcendent experience it would be, for so many people, to know that we share the earth with creatures who are spiritual as well as smart and emotional!

I have spent some time discussing altruism, empathy, and emotional sensitivity in apes, but no one should imagine apes are the Buddhas of the animal world; they are not taken up with quiet contemplation and compassionate practice. Gorilla males routinely kill infants when they take over a new group of females. Chimpanzees can be especially brutal: they plot against their rivals, ostracize their ex-friends, and murder their enemies. The cruelty inherent in chimpanzees' acts of physical violence is a horror to comprehend: when murdered by their fellows, chimpanzees die in terror and pain.

Perhaps humans would be unlikely to learn anything from a species that is invariably peaceful, a kind of paragon of animal virtue. (Does such an animal exist?) To define humans as a peaceful or compassionate species would be a sad joke. The cruelty of humans to other humans is as rampant as is human indifference to human suffering. Think of Darfur, Sarajevo, or Rwanda, or simply watch the evening news in New York, Chicago, or Los Angeles. That chimpanzees are capable of cruelty makes their acts of compassion, altruism, and empathy—and of meaningful attention to the natural wonders around them—far more meaningful to me.

No simple "ape nature" exists, either within an individual animal or within a species. If I needed any confirmation of this principle, so clear to me from watching bonobos and gorillas, I found it abundantly in the reports of fieldworkers and of researchers working with captive populations. Individual apes are radically affected by their environments—by the contours of their wild or captive habitats and by the dynamic interactions of the other apes who live with them. Furthermore, ape behavioral patterns differ appreciably across populations. Goodall's apes at Gombe groom each other frequently, but none uses the high-in-the-air handclasp position favored by their fellow chimpanzees across the country in Mahale. Gombe chimpanzees gobble up termites using fishing wands that they create, but do not use rocks to smash open the nuts lying all around them, as do their counterparts across the continent in Côte d'Ivoire.[19]

Populational differences in ape behaviors, beyond the basics of feeding or tool use or grooming, have scarcely been studied. Meanwhile, as I learned through my own ape-watching, *individual* differences can be fascinating and funny to watch, akin to the humor involved in recognizing idiosyncrasies in two siblings or two friends. Jane Dewar, together with her husband, Steuart, founded a gorilla sanctuary in Georgia called Gorilla Haven. In 2003, Dewar wrote about the gorilla Joe's idiosyncratic communication: "Now, 'most' gorillas show their annoyance by tightening their lips . . . It's quite a telltale sign of a gorilla's displeasure, and usually you get the hint really fast. Joe, on the other hand, has his own unique 'twist' to this gorilla behavior: he displays his lower teeth, which makes him look ridiculous and it's everything I can do not to laugh at him."[20]

Apes are the species most like us genetically and anatomically. Could it be that our closest living relatives reach heights of emotional relating and behavioral flexibility, and induce feelings of spiritual connection in us, more than do animals more distantly related to us?

ELEPHANTS

When Shirley and Jenny reunited, there wasn't a dry eye in the house.[21]

Everyone at the Elephant Sanctuary in Hohenwald, Tennessee, knew enough to expect *some* emotion that day in July 1999. Shirley, a fifty-one-year-old, 9,000-pound Asian elephant, was due to arrive by truck from a zoo in Louisiana. She was about to start a new life free of tight confinement, free of chains around her legs: a life in open land and green valleys, a life shared with other elephants. Captured in the wild at age five, Shirley had been brought to the United States to perform with a circus. This she did for about thirty years. Then she was taken to the zoo in Louisiana. She lived alone there, apart from all other elephants.

At the zoo, year after year, Shirley was unable to touch her trunk to another elephant's when she needed comfort, unable to trumpet her excitement or her nervousness together with one of her own kind. What did this isolation feel like to Shirley? We can't know for sure. We do know that elephants are highly attuned to their kin and their close companions. In the wild, the females and youngsters live in close-knit groups. The females rule, and their extended family members cluster around them; together they express their joys and sorrows. Just as chimpanzees and bonobos come together and separate in fission-fusion patterns, the subunits of elephant society part and reunite.

Perhaps, then, her time in the zoo felt to Shirley as it would feel if we were required to live apart from family and friends, indeed from all other humans. Surrounded by other species and subject to the schedules and desires of those in charge, the choices involved in eating, exercising, sleeping, and socializing would be beyond our control.

Luckily for Shirley, Solomon James came into her life. Companions at the zoo for more than two decades, Shirley and her caretaker Solomon shared many quiet and gentle moments together, the kind of moments that were few and far between for Shirley. Her circus days had not been easy. At age twenty-eight, Shirley's leg was fractured severely when another circus elephant attacked her. For reasons known only to the conscience of the circus owner, her broken leg was not set. As a result, the fracture caused continuing problems for

Shirley, and it was this same injury that later led zoo officials to quar-
antine Shirley away from other elephants.

Everyone who cared for her now believed that the Sanctuary, with
its thousands of acres and its promise of freedom, would make an
immeasurably better home for Shirley than the zoo. On the big day,
when Shirley arrived, she was welcomed immediately with delicious
foods, warm words, and a cool hosing-down shower. The first elephant she met was Tarra. Tellingly, this initial encounter is described
by the Sanctuary's elephant experts in terms of empathy: "Shirley
showed Tarra all her injuries . . . Tarra sympathetically inspected each
injury and the two elephants caressed each other with their trunks."[22]

Later that evening another elephant walked into the barn, near to
Shirley's stall. This was Jenny. Because Jenny seemed so intensely interested in Shirley, she was allowed to move into the stall next to Shirley's.
Now separated by a steel gate, the two elephants began to exhibit
behaviors both intense and unexpected. Carol Buckley, founder of the
Elephant Sanctuary, explained what happened at that moment and
over the next day: "Jenny wanted to get into the stall with Shirley des-
perately. She became agitated, banging on the gate and trying to climb
through over and over. After several minutes of touching and exploring each other, Shirley started to roar, and I mean *roar*—Jenny joined
in immediately. . . . We opened the gate and let them in together . . .
they are as one bonded physically together. . . . All day [July 7, 1999]
they moved side by side, and when Jenny lay down, Shirley straddled
her in the most obvious protective manner and shaded her body from
the sun and harm."

A photograph of Shirley straddling Jenny in this protective way
can be found on the Elephant Sanctuary website. Portions of the
reunion were shown on the television program *Nature*, in a film by
Alison Argo. When I saw the photo, my eyes filled with tears, and
when I watched the televised footage, I cried. I wasn't alone.
Filmmaker Argo remembers, "We were all in tears as they trumpeted
and rumbled and caressed each other with their trunks." The image of
Shirley waiting so very long to join with other elephants came into a

terrible focus for me, and I think for others, too, who witnessed Shirley's response to Jenny. In my case, the tears were as much of empathetic joy as sorrow, for Shirley is now freer than she has been since age five.

Why, though, did this emotional event occur in the first place? Why did Shirley and Jenny respond so strongly to each other, as compared to how they responded to other elephants? Shirley had found Tarra to be a pleasant companion, for example, but her response to Tarra was a gentle wave compared to a great tsunami with Jenny.

If you know anything about elephant behavior and intelligence, you've probably guessed that the answer lies in Shirley's past—and in Jenny's. With elephants, as with people, you need to know about the past in order to fully understand the present. Eventually the Sanctuary folks pieced together the story. For a single winter, some twenty years before, Shirley and Jenny had traveled together in the same circus. Already in her twenties at the time, Shirley was old enough to be Jenny's mother. Perhaps there were aspects of a mother-child relationship at work in that original closeness. Whatever had transpired between the two in the past, the incredible emotion of Shirley and Jenny's reunion in 1999 lends credence to the cliché that an elephant never forgets.

The more I learned about Shirley and Jenny, the more entranced I became with the Sanctuary's mission and accomplishments. A word of warning: the elephant bios and the live "elephantcam" on the Sanctuary's website are addictive! On the website, founder Buckley calls the Sanctuary a "sacred valley." Why sacred? I do not know Buckley's reason, but here's how the situation strikes me: When a place is created by one species, humans, in order to enable a sentient being of another species, elephants, to escape concrete and chains; to roam with friends new and old (of the same species) over green grass; to awaken other people to the emotional depth of elephants; and to attempt to make lives better for elephants everywhere, it's surely a special place, with feelings invoked that are larger than life.

There is something mysterious, something ineffable, in the

response of highly intelligent and emotional animals to death. We saw it in wild chimpanzees and in captive gorillas. There's something that binds us with animals who recognize a passing of this sort. Whether or not it's the passing of a soul is a question beyond my capacity to answer—but surely it is based on the recognition of loss, of a transition between total aliveness and then something different, something unalterably other.

Jenny died in October 2006, after a period of suffering from an undiagnosed illness. The lack of diagnosis did not stem from lack of effort, care, or expense; the Sanctuary went all out for Jenny, hand-feeding her at times. The story I want to tell now is not of her illness, but of her companions' response to Jenny's dying. Once again, Carol Buckley provides an eyewitness account:

> Last week Jenny could no longer engage in her normal foraging and migrating activity. She chose a beautiful forested wash area to lie down and rest until her time came to leave this world. Shirley, the closest thing to a mother that Jenny had known, stood protectively at Jenny's side, day and night, helping her to rise when Jenny shifted her weight to lie on her other side. Flanked by [elephants] Bunny and Tarra, they all seemed to know a serious change was occurring. On the day before her passing, Jenny engaged her sisters in the most profound chorus of rumbling as [humans] Carol and Scott stood witness and caressed Jenny, allowing the vibrations to penetrate their very souls. Everyone had accepted that Jenny was leaving, and it was obvious that this group song was an important part of Jenny's dying process, a process that excluded no one and drew her loved ones to a most intimate space with her. The joy-filled singing lasted for three hours. With each exhalation came a most relaxed and familiar rumble which drew Jenny's family deeper and deeper into the ritual, with Bunny adding a crescendo trumpet and Tarra chirping her excitement.

The day after Jenny's death, Buckley described the dying process as a peaceful one: "Shirley had moved away, painfully sensing that Jenny's death was very near; her sorrow was heavy. To lose Jenny for a second time was more than Shirley could bear. For the next few minutes Jenny uttered the baritone rich vibration with each exhale; it was not anguish, nor pain; she was calm and relaxed. . . . [Then] Jenny . . . was gone. Tarra and Bunny remained at Jenny's side throughout the night. Bunny remains even now, the following day, standing vigil, ever attentive to her dear sister, Jenny."

As she had not been left alone either in life or in dying, Jenny was not left alone in death. Elephant lives are lives shared. Intense greetings and intense partings happen among elephants living in the wild, too. This isn't to say an elephant is an elephant is an elephant. For most animals, rearing and environment will matter intensely, just as it did for the bonobo Kanzi in coming to learn so much language. In her book *The Elephant's Secret Sense,* researcher Caitlin O'Connell makes this point in a striking way. O'Connell works in Namibia and divides her time between two locations. In the Caprivi region, the elephants raid and eat farmers' crops. Understandably, to the community, these elephants are just pests. Tension and hostility marked the elephant-human relationships when O'Connell was there. Also, the Caprivi elephants, just like the Caprivi people, had suffered through Namibia's war for independence and all of its horrific costs. The elephants were no strangers to land mines, weapons, and poaching. Scientists have discovered that elephants suffer from post-traumatic stress disorder, just as humans do. Symptoms of elephant PTSD may include depression and hyperaggression, as reported by Gay Bradshaw, Allan Schore, and coauthors in the journal *Nature*.[23]

Compared with the Caprivi, Etosha National Park, particularly an area called the Mushara, is an oasis of serenity. Elephants there are relatively free of stress and their interactions mirror their more relaxed nature.

We should be as wary of reaching species-wide conclusions about elephants as we are about chimpanzees. Still, sometimes, wild elephants do express joyful feelings. And as any human knows, creatures who feel soaring joys also feel terrible griefs. When a wild elephant dies, she is unlikely to go unmourned.

Had I stumbled across the following account in a newspaper, I might have read it as a skeptic. Even now, trusting its authenticity and with a readiness to meet animals on their own terms, it surprises me. It was written by the prominent elephant expert Cynthia Moss, the same Cynthia I knew, a little, in Amboseli National Park when I studied baboons. In her book *Elephant Memories,* Moss first debunks the myth of an elephant graveyard. Contrary to legend, elephants don't flock to a certain location to die. Rather, hunters may kill large numbers at once, so that the bones are found in a cluster, in effect mimicking a graveyard. Moss then goes on to explain that elephants do seem to have what she calls a concept of death:

> It is probably the single strangest thing about them. Unlike other animals, elephants recognize one of their own carcasses or skeletons. Although they pay no attention to the remains of other species, they always react to the body of a dead elephant. . . . First they reach their trunks toward the body to smell it, and then they approach slowly and cautiously and begin to touch the bones, sometimes lifting them and turning them with their feet and trunks. They seem particularly interested in the head and tusks. They run their trunk tips along the tusks and lower jaw and feel in all the crevices and hollows in the skull. I would guess they are trying to recognize the individual.[24]

Moving as this description is, it remains abstract, akin to learning about human behavioral practices when a family is first confronted with the body of a loved one after a death; but the power of Moss's observations hits home in the concrete.

At Moss's camp in Amboseli, elephant jaws were left to dry out in the sun. I remember seeing them lying about, the few times I visited.

That way, the jaws bleach and become suitable for aging calibration. On one occasion, Moss writes, the jaw of a large, recently deceased adult female elephant had been lying in camp for three days. Moss had known the elephant who had died, and when that elephant's family passed through camp, she observed their actions. The family detoured right to the jaw and inspected it. Moss recalls, "One individual stayed for a long time after the others had gone, repeatedly feeling and stroking the jaw and turning it with his foot and trunk. He was the dead elephant's seven-year-old son, her youngest calf. I felt sure that he recognized it as his mother's."

Is this a concept of death, as Moss says? I don't know. The issues that capture my imagination are a little different. I have no doubt that the young calf missed his mother, and grieved for her. Did he also wonder where his mother had gone? Did he have any sense that the bones meant she was gone permanently? Can an elephant sense death, sense something about the transition between the alive and the not-alive? Can this calf's response to his mother's bones be considered a kind of ritual? Could there be such a thing as a religious ritual in non-human creatures?

During her fieldwork, Moss once found the carcass of a young female elephant. An elephant family encountered the body at the same time. Not only did these elephants feel and smell the body, they also did something more: they "began to kick at the ground around [the body], digging up the dirt and putting it on the body. A few others broke off branches and palm fronds and brought them back and placed them on the carcass." This astonishing behavior was interrupted when a ranger plane neared. The occupants of the plane were about to recover the tusks from the carcass, and in their zooming low in the air they so frightened the family that its members fled. Moss concludes, "I think if they had not been disturbed they would have nearly buried the body."[25]

For the moment, let's accept Moss's speculation that the elephants, had they not been scared off, would have partially buried the carcass. What does this mean? Could a burial of this sort be considered a ritual, and if so, could it be related to a sense of death, or even to a sense of the afterlife in the elephants? Much caution is required here. In *Evolving God*, I explored reasons that human ancestors who are totally disconnected from any sense of the afterlife or a religious sense might wish to bury a body. These alternate possibilities range from hygiene (the realization that a decaying corpse represents a threat to health), to respect or even worship of the ancestors. None of these explanations needs to involve any spiritual component. Of course, in human evolution or in human society at present, there may well be ritual of some kind, often relating to the afterlife, that does connect burial to religiosity. Where in this panoply of options would elephants' burial of a dead companion fall? Could the behavior relate to protection of the body of a loved one from potential or feared harm— or is it something more mysterious?

Perhaps these questions strike you as "out there," or "flaky," or an unwelcome sort of New Age spirituality, because we are asking them about a nonhuman animal—and a non-primate mammal at that. Have we left the realm of science in asking such questions? Paying attention to our relative ease or unease with these questions is itself an instructive process. The arena of animal emotion is wide open for scientific analysis, and as yet we have no answers to the most compelling questions. And some of those questions turn back on us: Do we feel a greater pull toward those animals, like apes and elephants, that mirror back to us, in their eyes or through their gestures and body movements, our own emotional worlds? Might we feel the greatest pull of all when we revel in the fun of shared joy or the comfort of shared sadness with our companion animals?

We have seen that both apes and elephants, in their interactions with other members of their species, express a good deal of emotional bonding and capacity for deep states of feeling. Some scientists think they may even see a glimmer of something more, of these animals understanding something about death or even larger forces that go beyond the everyday. Can the same be said of the animals we invite into our homes?

9

Dog and Cat (and Buffalo) Mysteries

•　　•　　•　　•

Little Dog's Rhapsody in the Night (Three)

He puts his cheek against mine
and makes small, expressive sounds.
And when I'm awake, or awake enough

he turns upside down, his four paws
*   in the air*
and his eyes dark and fervent.

Tell me you love me, he says.

Tell me again.

Could there be a sweeter arrangement? Over and over
he gets to ask it.
I get to tell.

—Mary Oliver

MATTHEW HURLEY TRAVELS OFTEN, for business. He and
his partner, Keith Lustig, have been together for so long, over
twenty years, that the unpredictabilities inherent in Matt's frequent

flying in and out of O'Hare Airport do not faze them anymore. When Matt is due to fly home, Keith has in his mind a general time of day that he's scheduled to land. Keith doesn't, though, pay much attention to a printed travel itinerary or exact arrival times. Air travel being what it is today, there's not much point.

One member of Matt and Keith's household *does* pay keen attention to Matt's travel schedule. Somehow, Riley the dog knows when Matt's plane has landed, and his behavior reliably communicates that knowledge to Keith.

About an hour before Matt's plane lands, Riley begins to "get hyperactive." "He runs and checks every window and starts pacing," says Keith. Then, inevitably, Riley calms down, and enters a second behavioral phase: "He goes to the window nearest the driveway, and puts his head on the sill, and waits for Matt's car." Soon after this sequence has unfolded, Keith inevitably gets a call from Matt: Matt's on the way home from the airport.

When I heard this story first from Matt, I was fascinated, especially because Matt was obviously a level-headed guy.[1] At that time, I knew in a vague kind of way about research that suggests some pets seem to know, somehow, what is going to happen before it happens. I was even aware of Rupert Sheldrake's book *Dogs That Know When Their Owners Are Coming Home* (but I hadn't read it). Despite being animal-oriented, I had not followed up on that line of thinking, because it struck me as uncomfortably close to invoking ESP in animals. I wanted no part of a movement that embraced psychic pets. Still, there was Matt, telling me about Riley.

When Matt and Keith and I talked together, I couldn't help feeling skeptical at first. Oh, come on, I said, isn't the explanation for Riley's behavior pretty obvious? Dogs are, after all, famous for "reading" the nonverbal signals of their human companions, so Riley must be picking up on subtle cues coming from Keith. While Keith may not have an exact landing time set in his conscious awareness, he nevertheless may begin to act in special ways in anticipation of his partner's return home. Perhaps he begins to prepare a special meal, or has an air of anticipation about him that Riley has figured out, through a type of trial-and-error associative learning over time, is linked in a general way with Matt's return home.

Keith and Matt were adamant that it was more than human "cueing." The best evidence for their assertion is that Riley behaves in the way I have described—hyperactive in phase one, calmly waiting at the window in phase two, all timed correctly—even when Matt's plane is quite late. In this case, if the cueing hypothesis were correct, Keith would become "anticipatory" hours before Riley's hyperactivity would kick in. In other words, if Riley had reacted merely to Keith's cues, he would have been running around too early in the day and then peering out the window at the driveway for an awfully long time! And this doesn't happen.

So how does Riley do it? Is there anything mystical going on here?

Sheldrake would say that Riley is telepathic. Between dogs and humans who have a close bond, there is "an invisible cord" that is "elastic: it can stretch and contract." This cord, Sheldrake believes, "connects dog and owner together when they are physically close to each other, and it continues to attach dog to owner even when they are hundreds of miles apart. Through this elastic connection, telepathic communication takes place."[2]

Does Sheldrake mean to suggest that dogs have ESP? The notion of animal ESP is not uncommon. It's embraced by proponents of the same kind of fuzzy New Age spirituality that I mentioned back when talking about shamanism. Pet-psychic shows air on TV, and books assure us that we can connect emotionally with our pets in the afterlife just as much as in this life. According to one such book,

On earth, animals understand what we say when we talk to them, but they do not communicate with us by using words. Of course, they do talk to us by barking, growling, hissing, purring, braying, howling, and so on. But on the Other Side, we can communicate with animals telepathically. At home when we want to communicate with our pets or animals, we simply need to think

what we want them to know, and they will respond in kind. This
is how animals communicate with one another as well.[3]

This description of animal communication is, plainly put, nonsense. In a way, I wish it weren't—if only, at home with my cats, I
could merely think *Jenna, if you hiss and spit at Pilar one more time, you
will be banished alone to the porch for an hour, which you do not like!* and
see Jenna's disposition magically improve. It's harmful, though, to
spread this kind of message. Animal behavior scientists spend time
and effort to observe, experiment with, and learn about the complex
and fascinating ways that animals really do communicate; much good
science goes missing when people insist on turning animals into
miniature psychics.

Yet there's a thriving market for the kind of approach exemplified
by Penelope Smith's *Animal Talk: Interspecies Telepathic Communication.*
According to a review posted on Amazon.com, Smith explains that
most animals welcome their owners taking them in for spay or neuter
surgery, because they, the animals themselves, are relieved to have their
sexual urges and reproductive cycles out of the way. And guess what
happened when Smith had to deal with an insect invasion inside her
kitchen cabinets? She discussed the situation with the bug leader.

I consulted Smith's website, http://www.animaltalk.net/, and I
just can't get on board. That is, I do not believe that Smith can channel the thoughts of cats and dogs around the world, or can commune
with insect pests. I cannot take seriously her claim that when pets "are
asked the right questions, they usually can recall their early experiences and even past lives." To suggest that she can channel animals'
thoughts, and that those animals have had past lives, is wildly different
from suggesting that people and their pets may find a way to attune so
closely to each other that the pets are able to predict their owner's
homecoming. Still, Smith's apparent popularity is a useful reminder of
how deeply some people wish to break through and communicate
with the animals around them.

Let's return to Rupert Sheldrake. Sheldrake is no Penelope Smith.

He is a scientist with a PhD in biochemistry from Cambridge University in England. Advanced science degree in hand, he nevertheless writes unabashedly, "I've been doing research on pets that suggests that many of them are psychic."[4] Is this nonsense, too? Can there be some scientific way to test for, rather than just to assert the existence of, telepathy in animals? The main concept Sheldrake uses is that of *morphic fields*. His explanation of this concept is lengthy and a bit opaque. The basic idea, as I understand it, is that when creatures live together, they are linked not only by social bonds but also by a kind of invisible force that "permit[s] a range of telepathic influences to pass from animal to animal, from person to person, or from person to companion animal."[5] This morphic field then operates to bind together individuals in a *psychic system*: a group of interrelated organisms, among which telepathy may occur.

Is Riley telepathic? Does he operate within a morphic field that encompasses the system of Matt and Keith and the other animals in their home? Scientifically it is hard to accept ideas like psychic pets or morphic fields. Yet it can be highly scientific to think in terms of systems: family systems, communication systems, emotional systems. Relationships within families are best understood as entities in their own right.[6] That Riley's behavior is embedded in a powerful family system seems to me a rich way to try to understand what's going on.

Riley, age six, is one of three dogs sharing a home with Matt and Keith. He's the biggest dog, and the youngest; his two older canine companions are the English setters Millie, age nine, and Spencer, age ten. In this family of five, there's a whole lot of communication going on and, significantly, much attunement. By that I mean that the humans and the dogs live in a thick web of attention, where they are acutely tuned in to one another's moods and signals.

Part of the unusual degree and quality of animal-human relating in Matt and Keith's home may stem from a condition of Spencer's. Spencer was born deaf. When Matt and Keith adopted him, their family system shifted. As Keith puts it, "Spencer's deafness has changed ways Matt and I communicate with the animals. It's been a catalyst to move our family to the next level." Keith got the idea to try out, with Spencer, gestures adapted from American Sign Language. He chose signals that are easy to make, and to see, to increase the chance of successful communication. And it works! Among the signals to which Spencer responds, Keith reports, are those for *outside, come, toy, good boy, bad boy,* and *no.* This is not the place for me to enter into an extended debate on the nature of canine comprehension of human gesture; the important point for our purposes is that between Keith and Spencer, the system works. And it has ripple effects.

Keith believes that because of his and Matt's communication with their animals, their own nonverbal communication has been enhanced. Once, Matt was in the basement, asleep after a medical procedure that left him with a severe headache. Around midnight he woke up, and immediately three words came into his head: "Check on Keith." These words were not uttered by a person—no one was there. *They just came into Matt's head.* Matt climbed up two flights of stairs, to find Keith relaxed and watching TV. Other than a little heartburn, Keith said, he felt just fine. Still, Matt had a sense that he should stay nearby. Within thirty seconds, Keith went into convulsions and collapsed. Fortunately, because Matt was right there to summon them, the paramedics arrived quickly. Keith recovered completely.

This isn't to say that the two men can intuit, at will, events in each other's lives when they are apart. After all, it's not Keith the person but Riley the dog who knows when Matt is going to arrive home. Keith takes pains to emphasize that animal and human sensory capabilities are not identical: "I think that communication occurs on a lot of different levels. Think of communication at the scale of sound waves, and humans only hear certain pitches. Dogs hear different pitches. Different species fall at different places on the continuum of communicative abilities. I just think that, oftentimes, animals are working on a different frequency. I wouldn't use the word 'psychic.'"

I cannot offer a credible explanation for how Riley knows when Matt will come home. I know it couldn't happen without nonverbal attunement, but I'm not sure that's enough. We've seen this situation before, the inability to completely explain animals' abilities: how do elephants recognize the jaws and skulls of their dead relatives? What are apes or birds doing, or thinking, when they approach the body of a dead companion and act in unusual ways? We don't know. Yet, just as with the elephant, ape, and bird behavior around death, there's no denying the reality of a phenomenon going on with pets who know what they "shouldn't" be able to know.

I have noted that Keith Lustig prefers not to adopt the word *psychic* in describing what happens in the human-dog system in which he lives. He does embrace the word *spiritual*, in the sense that being with animals brings him out of himself, out of a solely human-oriented view of the world. I cannot provide a scientific explanation for *how* Riley knows that Matt is about to come home—or a spiritual one. What I can do is explore a possible outcome of being with animals: people feel connected to something bigger, to something beyond the self. It might be God, it might be a force of consciousness in the universe beyond you or me, or it might just involve an enlarged (but nonreligious) sense of oneness with the world.

PETS IN THE FAMILY

Think for a moment of every person, every man, woman, and child, who lives in Italy (over 58 million people). Let your mind roam over the inhabitants of the small towns nestled in the northern Alps or on the volcanic soils of Sicily, to those living along the teeming streets of Rome or Florence. Or think instead of the combined populations of Kenya and Tanzania (almost 70 million people), all the people who work on farms in the Rift Valley or in the business centers of Nairobi and Dar es Salaam.

Then consider that the population of pet dogs in the United States (73 million) exceeds either total. The population of pet cats is considerably larger (90 million) than either. Pet birds, rabbits, ferrets, snakes, and an array of unusual and exotic creatures also share our homes. In short, Americans choose to live with pets in huge numbers, and Americans are not alone.

It would be a mistake no well-taught anthropology student would make, to assume that what goes for parts of North America, Europe, and the developed world goes for all parts of the world. For one thing, income enters the picture in a big way. When a family is poor and hungry, it may be impossible to share food and living space with pets. This fact affects not just the developing world, but, in times of economic downturn or crisis, people in all countries.

In families that keep pets, attachment between people and animals may be intense—something I know from experience.

There I stood, wiping the tears from my eyes, in the women's bathroom of a Petsmart store in Newport News, Virginia. We had just handed over our foster kitten, Alex, to a stranger.

How people can bring themselves to dump a tiny kitten out of their car and drive off, I don't know, but it happens a lot. That's how Alex came to us. My husband Charlie and I manage a colony of feral cats a few miles from our house. Every night, Charlie arrives there at dusk (I sometimes accompany him or fill in, but he's the steady one, day in and day out). At one point swollen to over twenty cats, the colony now houses eleven (in part because we built a spacious and leafy enclosure in our yard for most of the rest). The cats are well behaved, with shiny coats and sweet faces. When we show up, the trusting ones lean in for a stroke. Others approach only hesitantly, and the shyest ones run behind a bush. All the cats' eyes light up, though, and their bodies telegraph their anticipation. They know that with our arrival comes good food, clean water, and a kind word.

One night Charlie arrived at the colony to find a pathetic sight: a big-eyed, scraggly, flea-infested kitten on its own. We scooped him up easily; he was not afraid of people, a tipoff that he hadn't been born wild, but had been living with people for at least a little while. When the kitten tested negative at the vet for feline leukemia and FIV (the feline version of HIV), we brought him home.

We named him Alex, we fattened him up, and when he reached

the grand milestone of 2.5 pounds body weight, we had him neutered. Soon he raced through our front-yard to spring upon unsuspecting insects, to shoot out from the bushes down the driveway, and to tackle any waving grass or wagging finger that came his way. We couldn't keep Alex, though. With six cats in the house already (and more in the outdoor enclosure), we risked becoming one of those eccentric households you see on TV, where animal control officers zoom up and remove seventy-five cats and the local media catches it all on camera as neighbors shake their heads. So we carted Alex off to Petsmart on an Adoption Saturday. The woman who adopted Alex passed our screening questions and our home visit, and everyone ended the day happy.

So why did my tears fall? Because Alex wasn't a generic "rescue kitten" to me: he was *Alex*, an orange flash in our yard, a purring bundle in our arms. Like every cat I'd had since I was child, he was unique. The cats we care for indoors, and the feral cats we feed outdoors, each has its own quirks and endearing traits (and sometimes not-so-endearing ones). Together and singly, these cats give us enormous pleasure, as the cats who lived before them did in their day. (Lined up on our mantel are cherished small boxes, each labeled with a name like Horus or Mickey, that contain the ashes of animals we loved and still reminisce about today.) Obviously we're cat people. The right answer to "How big is your family?" is ten: three people, six cats, and one rabbit.

A rabbit? How did a rabbit get into the mix?

Years ago, when my daughter attended our local Montessori school, the primary classroom (home to the school's youngest children) housed a pet rabbit named Caramel. The teacher cared for Caramel, was kind to him, and allowed him to hop around now and then. Mostly, though, Caramel spent his time confined, looking out at the children—and, on weekends and holidays, entirely alone in a silent building, except when a volunteer arrived to offer food and water.

One summer we agreed to foster Caramel until school resumed. The family who had taken him the previous summer told us that

Caramel really didn't need much space: "All he does is sit there." They added, "We don't want him back."

Although I felt prickly on Caramel's behalf at hearing this, relating with him was more difficult than I expected. After years of looking straight on at apes' eyes and cats' eyes, rotated forward in the face as human eyes are, I was staring at an eye on one side of Caramel's head. This alone was flummoxing. And let's face it, looking into a rabbit's eye isn't the same as looking into an ape's. That spark of instant and intense recognition just doesn't flare. Rabbits neither gesture like apes nor purr like cats. How would I figure out *anything* about his mood, I wondered?

The best teachers were time and interest. I learned what any student of rabbits knows, that with alarm, rabbits thump their hind feet; the degree and quality of thumping indicates the degree of alarm. And Caramel had other ways to express his emotions, as does Oreo, the rabbit we live with now. When a rabbit is relaxing under a human's caress, the muscle tone in her body relaxes, too, and her eye (the one you can see!) very slightly closes. There's an overall sense of relaxed trust that is both visible and palpable to the touch.

The most rewarding times came when Caramel left his comfortable rug to hop onto the tile floor and make straight for where we sat in the den, reading or working or watching TV. When he did this, we knew he was choosing to be with us.

Slowly, then, Caramel and I learned to read each other, and to share rewarding times together. When, at eight years old, he died, the house didn't seem the same. Some time later, when we adopted Oreo, a smaller black-and-white rabbit, from the humane shelter, the house felt well balanced again. For us, *feng shui* seems to involve a 6:1 cat:rabbit ratio. (And no, the cats don't try to eat the rabbit!)

I have learned as much from living with "everyday animals" as from studying and writing about exotic ones. The joy of having a small creature knead its paws on my lap and purr loudly spills over into a feeling of shared love that radiates through my day.

I learned, too, to be suspicious of any sweeping statements about the nature of cats—their personalities, their behavior—as if one answer fits all. Are cats aloof and independent compared to dogs? The truth of this old chestnut depends very much on the cat and on the context (and, too, on the dog). Of course, no pet owner with intact senses would claim that dogs and cats have identical natures. I suspect that after a long day out of the house, Matt Hurley and Keith Lustig are greeted by their dogs with a version of full-body-quivering, tail-wagging enthusiasm that is duplicated by no cat in my house. Still, ask my husband, as his obsessively affectionate cat Jenna tangles herself in his legs as he tries to walk from the kitchen to the den, if his cat is "aloof and independent."

No longer do I see a cat and assume anything about its nature. When I meet a new cat nowadays, I look forward to understanding something about what makes her tick, and how we can create a space together where we may get to know each other. This process works even in the face of real unfamiliarity, as in dealing with Caramel the rabbit. And here's the thing: How could this happy anticipation about interacting with a new animal, or indeed encountering an individual from a species new to me, fail to affect my attitude and my inter-actions with people? Attempts to describe or judge individuals based on the supposed traits of some group to which they may belong—their religion, race, or sexual orientation—have never much appealed to me. Even so, I believe that an intuitive, or latent, understanding of individual humanity can be enhanced by insights gained from observ-ing and interacting with pets.

Could some version of "pet therapy" reform a racist, sexist, or homophobic person? Unlikely. Put a person given to racist stereotyp-ing ("Asians are smart; Latinos are lazy") in a room, give her the task of sussing out different personalities across the individual animals, and she is unlikely to become an equal-opportunity thinker. Yet, as we saw with Kanzi the bonobo and his language learning, or with elephants and the ability to empathize, when we really look at other creatures and see their unique abilities—both by species and by individual animals—it can rock our world.

Pets and their people are often astonishingly attuned to each other. This is the single greatest message I've gotten, whether talking with Matt Hurley and Keith Lustig, or listening to friends and neighbors talk about their pets, or consulting books about companion animals. People love to hear (and to tell) stories that focus on this attunement, and why not? We evolved to attend to other animals, to read their moods, to share our lives with them.

In a suburb outside the city of Hyderabad in India, a sixty-seven-year-old retired soldier and his sixty-three-year-old wife buried the dog they had lived with for thirteen years. A childless couple, the Madanrajs were very attached to Puppy, so much so that upon his death, they not only carried out a burial ritual for him, but also hosted a ceremonial feast in the dog's honor. After these memorials were completed, the husband and wife went into their bedroom and hanged themselves. In a suicide note, the couple described their grief over Puppy's death.[7]

What did Puppy give to the Madanraj husband and wife that they could not give to each other, or find in their network of friends? Mary Gordon, the American novelist, calls dogs "dread absorbers."[8] Maybe that's what Puppy could do for this Indian couple—bring them out of their human worries, their human anxieties, their human sadness in a way that no person could manage. The degree of attunement the Madanrajs shared with Puppy must have been great; yet it was also extreme. An emotionally healthy attunement does not end in suicide. Yet this story is extreme only in its outcome. Grieving over a pet, and memorializing a pet's death, are far from uncommon.

When Hurricane Katrina hit the Gulf region of the United States in August 2005, more than 1,000 people died. How many pets died has never, to my knowledge, been calculated precisely. We know that thousands of dogs, cats, and other animals were stranded along the Gulf Coast. I remember seeing heart-wrenching images of lost and bewildered animals on television and in weekly magazines, including a dog, appeal in its eyes, standing waist-deep in floodwaters outside a home destroyed by Katrina.

For New Orleans itself, we do have statistics: the Louisiana Society for the Prevention of Cruelty to Animals, which operates the city's animal shelter, estimates that 70,000 pets remained in the city during Katrina. About 15,000 were rescued, but only about 20 percent of that number were reunited with their owners.[9] Whatever the overall grim statistics, the tide of grief felt in the wake of these deaths and separations in Louisiana and Mississippi was surely enormous. A number of the human deaths can be understood as stemming from a refusal to abandon pets to the floodwaters.[10] Stranded on rooftops or on a top floor of their home with a cat or dog, and learning from rescuers that animals were barred from the hurricane shelters to which they themselves would be taken, some people simply would not go. News reports suggest that the problem existed on a large scale. And, tragically, not everyone who made the decision to stay made it out.

Even the Red Cross adhered to this no-pets policy for its shelters, as its hurricane checklist makes clear: "Make a plan and prepare to evacuate. Plan your evacuation route by using maps and identifying alternative routes. Pets should not be left behind, but understand that only service animals are permitted in shelters. Plan how you will care for your pets and bring extra food, water and supplies for them."[11]

During Katrina, this policy led to a desperate, sometimes fatal, decision by people to stay at home; just as they would not leave a child alone to face the floodwaters, or slow starvation, they did not leave a cat or dog. You just don't abandon a family member.

A pet's death, no matter the circumstances, is often a devastating event, one that is more and more, these days, grieved openly. Some people arrange for elaborate burials, as a burgeoning pet-memorial business attests. At Pets at Peace in Toronto, Canada, one pet lover's wishes were carried out when a black cat was arranged in death on a bed, in curled position; red beads, white flower petals, and baby's-breath flowers surrounded the cat's body. At a similar business in nearby Perkinsfield, a six-foot iguana was laid out, its head on a pillow, in a blue-padded mahogany coffin.

Anyone with access to a computer can rapidly locate an array of services ranging from home burial to cemetery burial to cremation of pets. In my own house, as I have mentioned, a small corner is devoted to the ashes of our past pets. When we lost one cat to cancer, our veterinary specialist sent us a poem that we now realize is famous among animal lovers. "The Rainbow Bridge" may not stand as a great contribution to literature, but it has consoled us, and many others, in sorrowful hours. The poem describes a bridge that pets cross upon death, leading to a place where health and vigor are restored, and the animals live in comfort and happiness.

> *The animals are happy and content, except for one small thing; they each miss someone very special to them, who had to be left behind.*

> *They all run and play together, but the day comes when one suddenly stops and looks into the distance. His bright eyes are intent. His eager body quivers. Suddenly he begins to run from the group, flying over the green grass, his legs carrying him faster and faster.*

> *You have been spotted, and when you and your special friend finally meet, you cling together in joyous reunion, never to be parted again.*

My response to this poem has nothing to do with a literal belief in an afterlife shared with animals, but rather with the acknowledgment that an animal has shared with me a unique one-on-one relationship that changed both of us.[12] For others, the poem may represent a wish for an embodied spiritual reunion with pets of the past, ranging back many decades. In irreverent moments, I can't help but wonder what it would be like to experience the sight of cats from my childhood romping toward me, only to spy the rival cats from my adulthood coming forward from another angle!

Poems like this, and fond memories of happy times, are shared in pet-bereavement support groups. As with any such group, the idea is that others who have gone through something similar will understand. Those who aren't "animal people" often don't understand how terrible a pet's loss can feel. Cheri Lieberman, leader of a pet bereavement group in Massachusetts, noted in a newspaper interview how off the mark is a response too commonly given: "People say, 'Get another dog,' and it's not that simple, because the pet that you've lost has a personality and you have a relationship with that pet. You can't take that one out and put another one in."[13]

And here is the crux of it: the grief felt at a pet's death is an index of the joy shared with that pet. Full circle, we're back again to attunement, because attunement *is* joy. What a wide range of animals evoke feelings of joy and connection beyond the self! Cats and dogs are the power pair of pets, yes, but everyone from bird lovers to bison lovers has stories to tell. And the stories show not only the fact of animal-human connection but the power of transformation that may be contained within that connection.

Anna Autilio, a high school senior who volunteers at the Raptor Trust in central New Jersey to help injured birds, and who intends to study zoology in college, described what happens when she interacts with her pet birds. "When I exchange *pretty birdie*'s and *I love you*'s with my parakeet Tiki, I feel loved and wanted by something that isn't human, which is very different from the human experience." She continued, "I feel like a part of his life, or a part of his world. I also feel loved and wanted when Pete gets frightened on my shoulder, and backs up into my neck, under my jaw where he's a little protected. It feels amazing to know he's comfortable there."

Whether indoors with her birds or outdoors watching a stag, Anna describes a feeling familiar to many of us: the animal is sharing its world with you then, for that moment, just you. It is as if we enter a special place when we relate with animals: not a physical location, but a state of being. Or it may be more accurate to say we *create* a special place. To create a special place with an animal requires a readiness to connect, to pay attention to all the little idiosyncrasies of interacting with this animal versus that one. As with the pet-bereavement support groups, a person tends either to "get" being in that place with animals—or not.

Not all that many people "got it" when Roger Brooks and Veryl

Goodnight poured time, effort, and money into relating with an animal at the opposite end of the size continuum from Anna Autilio's birds. Roger and Veryl, for a number of reasons involving Veryl's family history and her work as an artist, became keenly interested in buffalo. (I use the terms *buffalo* and *bison* to refer to the same animal.) They lived just outside Santa Fe, New Mexico, but knew a friend with a bison ranch in Idaho. In spring 2000, when Veryl wanted a model for a bison sculpture she was creating, they adopted a baby bison who had become separated from his mother. The baby needed TLC, and lots of it. Thus began the story of Charlie, *A Buffalo in the House.*[14]

When Roger and Veryl met him, Charlie was a red-colored, dog-sized, week-old baby. In the years the three shared, Charlie grew to 1,800 pounds and had the utterly unique shape of a mature bison, with massive head and shoulders and tapered body. Yet the story to be told here is not about shaggy coats and big animal bodies, it's about a wild creature coming to care for, and trust, two humans. First and foremost, it's the story of Roger and Charlie. Although Veryl is a key participant, there can be no doubt that the heart of the matter lies with the relationship between Roger and Charlie. It was, as R. D. Rosen puts it, a relationship "unlike any man-buffalo relationship in history."[15]

As a young calf, Charlie inhabited the house. He walked at will through the rooms, and guzzled gallons of milk held for him by Roger or Veryl. He was as much a pet at this point as were the household dogs. When Charlie began to bulk up—he packed on two pounds per day—he was at first discouraged from too much in-house time, and then banned from the house entirely. In his stall and around Roger's property, he enjoyed space to roam and fresh air. A high point of Charlie's day involved walks with Roger in the foothills of the Sangre de Cristo Mountains; buffalo and man threaded their way through stands of piñon and assorted other trees.

Before long, Roger and Veryl concluded that Charlie should be encouraged to discover his true nature, his bison nature, rather than continuing to live only with people. Two hours away, near Taos, was another buffalo ranch. There, inside a steel-fenced pen, Charlie was introduced to two other buffalo. Human hopes for the encounter ran high, but this reunion with his own kind caused no rejoicing on Charlie's part. After all, he could remember life only with humans, and from humans he had known only comfort and kindness. His closest companion was Roger. He showed no signs of interest in the bison. In fact, it was worse than that. Charlie had no clear notion, it seemed, that he *was* a bison. Still, Roger and Veryl left him in the pen, their sadness at leaving him mixed with hope for his future.

During that first night, apparently spooked by his proximity to the hulking unfamiliar creatures near him, Charlie ran headlong into the steel fence surrounding the pen. No person witnessed this accident, but it must have involved a forceful impact; the next morning, Charlie lay injured on the ground, near the fence. He had sustained severe neurological trauma, and it wasn't clear whether he would walk again, or even survive. Surgery, followed by intensive rehabilitation, was his best chance.

As quickly as they could drive the miles between Santa Fe and Taos, Roger and Veryl were at Roger's side. They arranged for surgery at a university's veterinary medicine division and pitched in, after the operation was over, to move Charlie's limbs in patterned rhythms, trying to induce his legs to work again. It was a slow and uncertain (not to mention expensive) road back, and while it's accurate to say that his body never fully recovered, Charlie fared better than anyone had expected.

When Charlie was relatively stable on his feet, Roger decided to bring him back home to Santa Fe. The pair even resumed their walks together. Here is where the words *relatively stable* assume their significance: "It wasn't like the old days," writes Rosen. "Charlie walked crookedly, his hindquarters twisted to the left. He still hoisted his hind right leg forward in an arc." When Roger noticed that Charlie might be about to cave in and fall, "he'd lean his hip hard against [Charlie] with all his strength; it was often enough to stabilize Charlie's [then] five hundred pounds."[16] When it wasn't enough, and Charlie fell,

Roger would purposefully look away. Roger not only believed with his heart, but also saw with his eyes, that the bison felt depressed and embarrassed when he tumbled to the ground. "Now he was hesitant to look Roger or Veryl in the eye. His grunts had become low grumbles."[17]

By this point, Roger was able to enter Charlie's world; he understood the subtleties of Charlie's physical condition and, more important, his emotional one. For his part, Charlie responded differently to Roger than to any other being: the relating was mutual and affectionate—though occasionally laced with signs of Charlie's emerging dominance. The walks themselves were the best evidence of the fact that attunement trumped dominance almost every time: "[W]ith some regularity, when Charlie was not in a mischievous mood, Roger could now get him to come just by calling, and could get him to go in a certain direction just by pointing." As Rosen emphasizes, this is far from a natural situation for a large wild animal. "This degree of influence may be no great accomplishment with a dog, but it's a rare cat who'll stand for it, and you had to see it to believe it in the case of a buffalo."[18]

Through all the ups and downs, the joys and sorrows, shared in this family of three, the attunement between Charlie and Roger stands out as the most remarkable aspect. That a bison would choose to respond to, and care for, a human as Charlie did with Roger attests to the power of the rearing environment. "Normal" bison don't do what Charlie did—just as "normal" apes don't do what Kanzi the bonobo did in language learning. But what was "abnormal" wasn't Charlie's genes or his biology, it was his environment. Charlie was transformed by the actions of others around him, just as Kanzi was. Charlie was surrounded by love and attention and also, it must be said, by a man who, being as human as any of us, sometimes made mistakes with him. At times I felt ill at ease with the degree and quality of human intervention in Charlie's life. That feeling is no doubt colored by the story's outcome. At age three, Charlie died of the bovine form of bacterial pneumonia.

By the time of Charlie's death, Roger had been spending an average of one hour each day with Charlie; the two had not been on a walk together for a full year. In some ways, Charlie had been a pet. Certainly he started out as a pet, fed and raised inside the house. Still, throughout his life, he remained a partly wild and partly domesticated creature, a hybrid ambiguity who raises moral questions for the rest of us. Was Charlie's life better or worse for Roger's keeping him away from other bison? Would the outcome have been different—no head-long rush into the steel fence, which set off a cascade of other events that ended in Charlie's premature death—had he been introduced to others of his kind more gradually? What if Charlie had been put in the pen for only a few hours at a time, at first, and allowed to be on his own (with visits from Roger) at night? Was the failure to anticipate Charlie's response to other bison a fatal failure of attunement?

These questions may be fair, but it's easy to second-guess Roger and Veryl. They responded, day by day, not to an abstract moral problem, but to a flesh-and-blood creature they loved fiercely, an animal they refused to abandon even when sending him elsewhere would have made their own lives markedly easier. In any case, Roger does a thorough job on his own of second-guessing his actions with Charlie, as is apparent in *A Buffalo in the House*. The full story is there for anyone who wants it, a tale of two creatures who come out of themselves to create a special space together. Roger's world opened up through Charlie; he now devotes a considerable part of his time to helping wild buffalo.

MORE ON UNEXPLAINED POWERS

The attunement that Roger and Charlie achieved together was hard-won. Unprecedented and extraordinary as it was, it was neither para-normal nor psychic in nature; it can be grasped by knowing something about the force of love acting on a young animal's development, and by knowing something about Roger's persistence and sensitivity. But the world of pets, people, and beyond-the-self transformation embraces a kind of human-animal attunement that strays into an area not as easily comprehensible, an area that flirts with the mystical. It includes the power of some animals to divine something that they shouldn't, by all rights, be able to know, just as Matthew Hurley and Keith Lustig's dog shouldn't have been able to figure out when Matt was due home.

With his research on what dogs know, Rupert Sheldrake provides a touchstone for exploring this special attunement. Telephone surveys, anecdotal reports, and videotaped experiments all lead Sheldrake to the same conclusion: dogs are capable of long-distance telepathic communication.

Sheldrake and his associates telephoned, at random, 1,200 households scattered across two locations in the United States and two in Britain. Questions asked were whether pets seemed to get "agitated" before a family member arrived home, and if the answer to this was "yes," how long before the arrival. On average across the four locations, 55 percent of dogs were reported to anticipate their owners' return, most within ten minutes of arrival. Judging from this survey alone, it's about an even bet whether any given dog will anticipate a human's return home. Riley, Matt and Keith's dog, is no lone genius, then, but not every can dog do what Riley can do. When dogs don't anticipate, writes Sheldrake, the bond between dog and human may be weak, or perhaps that dog is just plain lacking in the ability in question.[19] As I have stressed, dogs vary greatly in a number of ways, and it's not surprising that some dogs either are innately limited in their ability to anticipate, or have not been raised in a way that would allow their inherent sensitivities for anticipatory behavior to emerge.

The mere recording of pet owners' claims over the telephone does not amount to science. In-depth interviews may be more revealing. One British man reported the events that unfolded when he minded his sister-in-law's dog, as the sister-in-law and the man's wife shopped together in a town fourteen miles away. At 4:45 p.m., the dog walked to a window and sat there, but quite quickly returned to a relaxed position on the carpet. A half hour later it again became attentive and anxious. As it turned out, the two women had first started to leave the shopping area at 4:45, but changed their minds and stayed an extra half hour. The dog's behavior precisely mapped onto his owner's intentions! (Importantly, the dog-minder had not been told by telephone of the women's intentions.)

How eerie to think that some dogs respond to a person's thoughts! Such an idea surely falls outside the bounds of conventional science. It's an uncomfortable idea for me. Yet consider this experiment, videotaped by Sheldrake. A woman named Pam Smart, who lived in Britain with her parents, communicated with Sheldrake about her dog Jaytee's anticipatory behavior. One thing led to another and in 1994, Sheldrake's team set up two cameras in order to pin down what was happening. One camera made a continuous recording of Jaytee's behavior inside Pam's parents' house, while Pam was away; a second camera tracked Pam's whereabouts. Pam was instructed, at the last minute and by Sheldrake's associate, when to set off to return home. She neither knew the time in advance nor communicated it, once told to her, to her parents back home.

A split-screen viewing of the results from both cameras is revealing:

> To start with, Jaytee is, as usual, lying at Mrs. Smart's feet. Pam is then told that it is time to return, and almost immediately Jaytee shows signs of alertness, with his ears pricked. Eleven seconds after Pam has been told to go home, while she is walking toward the taxi stand, Jaytee gets up, walks to the window, and stands there expectantly. He remains at the window for the entire duration of Pam's return journey.[20]

Eleven seconds! The dog's response to Pam's movements is rapid. Still, from these results to a claim of dog telepathy is a big leap indeed, isn't it? Couldn't the cause of Jaytee's alertness have been a passing mail carrier or some other pedestrian, someone completely unconnected with Pam? Couldn't the dog's stationing himself at the window be explained by an enthusiasm for, say, watching squirrels dart around the yard at that hour? Correlation isn't cause, as we all know. In other words, we know that Jaytee did these things, but we cannot know why he did them.

Here's the kicker, though. The initial videotaped results led to the making of thirty more tapes, in two different years, that document Pam and Jaytee's "elastic cord" communication. Repeatedly the tapes show that Jaytee's behavior is explicable neither by routine patterns on Pam's part nor by cueing on the part of her parents. Is it credible to explain away the consistent outcome in Jaytee's behavior time and time again?[21] Just as Charlie the buffalo and Roger the person were highly attuned to each other on their walks, so the same can be said for Pam and Jaytee in their shared lives, and in this case the attunement extends to Jaytee's ability to tap into Pam's intentions at a distance.

As noted earlier, Sheldrake explains behavior such as that captured on the Pam-and-Jaytee tapes by the existence of morphic fields. I'm unconvinced by his idea of psychic connections via invisible forces, but I am convinced that the phenomenon of anticipatory behavior by pets is real. In short, I cannot explain it except to acknowledge the power of connection and transformation between two creatures. I don't mean that the dogs have some kind of sacred knowledge or spiritually rooted mystic ability. Rather, I mean that the anticipatory behavior in question emerges from that special space where people and animals may relate with keen attunement. For some people, this place is a spiritual one.

Cat lovers may be wondering why all the fuss involves what *dogs* know. In fact, it doesn't—in his random telephone survey, Sheldrake discovered that 30 percent of cats in the United States and Britain were said to anticipate their owners' return. Cat-oriented interviews can be every bit as intriguing as dog-oriented ones. One amusing story tells of a student who used a cat's behavior as an early-warning system when he hosted unauthorized parties at the island guesthouse where he was staying and working! The cat, a Persian, would anticipate his owner's return by about twenty minutes, no matter the hour or the method (car, boat, etc.). The sneaky student thus gained enough time to end the festivities before getting in trouble with his landlord and boss.[22]

Anticipatory behavior is not the only measure of attunement between cats and people. That cats (and dogs) are acutely sensitive to changes in their owners' physical conditions or moods is a favorite topic of conversation among people in my animal-oriented circle. In 1995, my friend and fellow ape-watcher Joanne Tanner had a wisdom tooth extracted. She bled rather a lot, and was told by her dentist to keep her mouth closed, with gauze pressed against the extraction area. Joanne returned home, and lay down on the couch. Right off, her cat Muffin approached. This behavior in itself was not unusual; Muffin often lies on Joanne's stomach. But there was an added element this time. Though there was no external indication of trauma to her cheek (the gauze, for example, did not puff out her cheek), on this day Muffin "put first one paw, then both, on *precisely* the spot where the tooth was extracted, and settled in and purred and purred. My position got uncomfortable and I moved my head around—she brought her paws back to the very same spot."[23]

Joanne suggests that a change in temperature at the affected area may explain Muffin's behavior. If so, perhaps temperature change had something to do with my friend Karen Flowe's experience with her cat Carson. After suffering with pain and stiffness in her shoulder, Karen was given a professional—and quite vigorous—massage. When she returned home, still in some pain, she stretched out flat on her back on the couch. She drifted off to sleep. When Karen awoke, she felt a warm sensation. Her cat Carson had positioned herself across the affected area in much the way one would position a heating pad. Carson's hind feet were on the couch, and her stomach was right over the painful area; with her front paws, Carson was kneading the front of Karen's shoulder. "She seemed," Karen told me, "to be clearly seeking to minister to my needs. I felt loved, comforted, and oddly understood."

Even if Joanne Tanner and Karen Flowe's cats operated from some kind of temperature-detection ability, there's no denying they were exquisitely attuned to their people. I've experienced this phenomenon myself. One day, I went to my eye doctor and heard distressing news: the blurriness I had noted in one eye was caused by post-vitreous detachment. PVD occurs when the eye's gel or vitreous tugs on the retina, causing cloudy vision and floaters. Sometimes this condition progresses to a retinal tear or full retinal detachment, a serious development. Statistically, PVD is *unlikely* to lead to real trouble of this nature, but in my case I had cause for worry: eight years before, the retina in my other eye had detached. After emergency surgery to put in a metal buckle and a gas bubble and to extract the vitreous, I had had to keep my head down for fifty minutes out of every hour, for ten days straight, in order for the cure to "take." That was, to put it mildly, a memorable span of time.

When I received the new diagnosis, I reassured myself with statistics: the progression of PVD to retinal detachment was probably not going to happen. My mind may have accepted this statistical evaluation, but my body did not. Within an hour, my back went into painful spasms. When I lay down on the floor to try to ease the pain, my cat Pilar came to sit with me. She did not lie on the affected area (I was lying on my back), but she responded right away. Was this because I had adopted an atypical position, in an atypical location? Maybe, but her response was more comforting than curious, and it didn't extend to the other curiosity-prone creatures in the house. The cat I am closest with was the single cat to respond to me.

Perhaps, then, these acute sensitivities at work between animals and humans during up-close interactions may stretch to situations where owners are physically at a remove. Touching or lying on an affected area or comforting a sick person would then exist on a continuum with anticipating an owner's return.

One example of a cat's powers stretches beyond anything I've encountered otherwise, in life or in conversation. In 2007 a cat captured the headlines for a remarkable ability—this time, not to know when a person was to return home, but when a person was about to die. Writing in the *New England Journal of Medicine*, hardly a hotbed of New Age woo-woo, Dr. David Dosa described the behavior of a gray-and-white cat called Oscar.

Oscar had been living in a nursing-home facility in Rhode Island for a year and a half, since the age of six months. The staff noticed that Oscar made rounds on the facility's dementia unit, moving from one patient's room to another. He closely observed and even smelled the patients. Gradually a pattern emerged: when Oscar settled in next to an aged patient, right on the bed curled against her body, the person would die within four hours.

Oscar's ability led to a new practice at the nursing home: when he curls against a patient, the nursing home summons the family. At the time Oscar came to media attention, he had correctly predicted death twenty-five times. Almost always, the patients themselves were too ill to be aware of Oscar's presence, so the cat's predictive powers did not distress them. Many of the patients' families expressed gratitude for a little extra time to prepare themselves for their impending loss—and for the companionship that Oscar offers. I cannot help but think in poetic terms: Oscar acts as a sort of animal guide to the Rainbow Bridge, a guide who eases, just a little bit, the crushing loss when a loved one dies.

What potential lies in this kind of attunement between animals and people! It is possible, even probable, that Oscar's behavior is explained by something completely apart from compassion: he may sniff out an odor unnoticed by humans, a smell that accompanies a chemical reaction as the body goes through the last stages of dying. This does not matter. Behavior like Oscar's with dying people, or Charlie the buffalo's with his human friend Roger, or any of the other examples in this book that involve animal-human relating, can unlock *our* compassion, *our* best selves. When we experience something special in the space where animals and humans relate, our reverence for the world may become supercharged. If we are religious, we may feel moved to pray in order to connect with our God. If we are not religious, we may feel heightened "oneness" with the natural world. When the biologist Ursula Goodenough contemplates the beauty of all life, she embraces a credo of continuation:

> For me, the existence of all this complexity and awareness and intent and beauty, and my ability to apprehend it, serves as the ultimate meaning and the ultimate value. The continuation of life reaches around, grabs its own tail, and forms a sacred circle that requires no further justification, no Creator, no superordinate meaning of meaning, no purpose other than that the continuation continues until the sun collapses or the final meteor collides.[24]

Maybe a Rainbow Bridge leads us, all of us united without regard to faith, agnosticism, or atheism, not to an afterlife full of carefree pets reunited with their owners. Maybe it leads us instead, through our connection with other animals, back into this present life of ours, toward a will to care, to comfort, and to act to reduce suffering in animals and people wherever it is found.

10

Finding Compassion

●   ●   ●   ●

When we care for animals, we ache for them when they
suffer. Solace may be found in the animal beauty of
Thomas Mangelsen's photography:
http://www.mangelsen.com

MISS OINKERS WAS an adorable piglet when a family in
Charleston, West Virginia, adopted her. For a while, potbellied
pigs like Miss Oinkers were all the rage. But like most fads, this one
faded over time. And of course, cute piglets grow up to be large,
difficult-to-manage pigs. Her family soon became indifferent to Miss
Oinkers—and worse. For seven long years they confined her to a dark,
cold basement. Sparing almost no time for her, they overfed her so
dreadfully that she swelled to 400 pounds and more.

Finally the family wanted nothing more to do with Miss Oinkers.
A resourceful and kind person, aware of the pig's plight, built a cage
around Miss Oinkers, removed the basement door so that she could
be carried through it, and drove for six hours in order to bring her
to PIGS Sanctuary near Shepherdstown, West Virginia. A grim sight
greeted the PIGS staff upon Miss Oinkers's arrival: "It was one of
the worst cases of neglect we have seen. Since she was locked in a

basement and deprived of sunlight, she never shed her coat. Her hair was almost a foot long in some places. Her skin was infected, dry, and cracked—brought out by poor nutrition, and her negligent environment. Rolls of fat covered her eyes, prevented her from seeing. She literally walked on her very rotund stomach that dragged on the ground."[1]

The road to recovery for this pig is long, but a PIGS update indicates that she is able to see again; she has lost over one-quarter of her body weight; and she is beginning to trust the staff.

Miss Oinkers is just a drop in the bucket of indifference and cruelty to animals. Here is the flip side of that special space created when people invite animals into their families. Indifference and cruelty create a different kind of space, one that feels like hell to the animals. Every single story represents a broken life—a pet abandoned, a farm animal made to suffer in crowded conditions, an endangered wild animal hunted down in order to feed the craving of wealthy diners for upscale meats. It's a reality that is part of the being-with-animals story, but it is a reality that is laden with hope: hope for change *because* of humans' robust evolutionary history of emotional transformation with animals.

INTO THE WILD

According to TV's Travel Channel, among all national parks in the United States, Yellowstone ranks third for wildlife-viewing, and nearby Grand Teton National Park reigns at the top spot. For me, these parks vie for alpha position in a more specific category: up-close and thrilling bison-viewing.

In the summer of 2007, I hadn't yet read about Charlie, the buffalo raised by Roger and Veryl Brooks, or spent much time thinking about buffalo at all. I'd admired buffalo a few years ago in a drive-through safari park in Oklahoma, but in Wyoming it was different. It was my first visit to that part of the West. The infinite skies canopied mile after mile of dizzying openness: plains rolled right up to glacier-studded mountains, and valleys plunged down to rivers snaking along with chilly water. Herds of elk grazed on grass. A moose rested quietly here, a coyote trotted by over there. And I experienced a memory held deep within my body, indeed within my muscles: how wonderful it is to still oneself and enter the world of other creatures. I thought back to my time spent in Kenya, observing monkeys. Just as in Kenya, in Yellowstone the very air takes on the tension of animals poised on the knife edge: should they stay and look back at you (and thus open up the space for relating), or flee?

The variety of animals in Yellowstone is amazing in itself. Tim Cahill refers to the area as "the Serengeti Plain of North America" because "all the creatures of the plain and the mountains converge in this landscape of marvels," and he's right.[2] On our very first night, with the help of a friendly network of bear-watchers armed with spotting scopes, we spied a grizzly in the medium distance. Soon the bear could be seen with the naked eye, moving his great bulk with grace down toward the river. The next day, as we drove to view steaming geysers and rainbow-colored hot springs, we gaped as a huge grizzly padded right past our car, his big, sloping head and shoulders pointed forward in a calm walk on an easy-to-navigate surface.

It was in the car, too, that I sighted my first wild bison. This was before we had entered Yellowstone, as we were still driving in from Jackson, Wyoming. The road was lined with trees and fences that half-shielded from view the big ranches on either side. When I happened to glance left, I glimpsed a big shaggy brown shape making its way along, behind the fence. It sounds ridiculous now, like a poorly acted scene in a movie, but at that moment I could not speak for excitement. I gripped my husband Charlie's shoulder (he still teases me about the aching aftermath of that iron grasp) and then, regaining myself, screeched "Bison!" so loudly it brought my teenage daughter out of her iPod-induced reverie.

We stopped the car. Emboldened by the fence between us, but still keeping a respectful distance so as not to stress the animal, we watched and photographed the bison. What riveted me so, watching this animal? Partly it was the sight before me, the solid body eating its way calmly through the world. Here was a lone bull: when would he meet up with a female? (It was August, the rutting season for bison.) How would he act in a social group? There was so much to wonder about. Also, we were gazing at a living symbol of our country; the history of the United States is entwined with the history of the buffalo.

When another car stopped and two people joined us to view the buffalo, we departed. An hour later, inside Yellowstone itself, near a Snake River overlook, we encountered the same people again. They told us that after we had departed, the bison had smashed through the fence. Had he been provoked? What caused the shift from quiet creature to fence-smasher? We would never know.

The lone bull was a harbinger of what Yellowstone would offer us. On our first night, we drove out to Hayden Valley. Bison everywhere! A herd spilled onto both sides of the road, grazing on flat land and small hillocks. We knew better than to get out of our vehicle. The book *Death in Yellowstone* tells what happens when people are foolhardy enough to believe they can outwit or outrun an annoyed bison.[3]

Bison-viewing from a vehicle may seem a little bit on the sterile side: after all, how close can a person feel to nature from inside chrome and steel? It all depends on one's perspective. That night we rolled down our windows, stayed still, and listened. Bison are surprisingly noisy. Males grunted as they mate-guarded their females; calves galumphed around in play, on still-shaky legs, only to return to their mothers and nurse. Unfolding before us were lives full of social intricacies, of close bonds, and of rivalries, too.

Behind the apparent serenity of the Yellowstone bison lies a monstrous history. The story of bison in the United States is the story of Miss Oinkers the pig writ large: of self-indulgence (or even greed) vying with indifference. Until around 1860, bison outnumbered people in the United States. The herds comprised such numbers that we have no example from our own lifetimes from which to draw a proper parallel. Imagine the sight described by N. D. Rosen:

> The country was teeming with buffalo, herds of them hundreds of thousands strong, a vast shaggy carpet of buffalo miles long and miles wide. On first seeing them, men were apt to rub their eyes and question their sanity. There were places where the buffalo weren't just on the land; they seemed to *be* the land. If you couldn't see them, you could hear them coming, well in advance, an apocalypse on hooves. Sometimes it took days for them to run past a fixed point.[4]

A map of the "maximum bison range" in North America has me rubbing my eyes, too. Somehow I'd always thought of the buffalo as Western creatures, at home on the Great Plains, blanketing certain states and parts of Canada, but strangers to the East Coast or the American South. According to the map, though, bison once roamed through Pennsylvania, into my home state of New Jersey and my adopted state of Virginia, and south through Georgia and Texas into Mexico.

How many buffalo once walked freely over these lands? Historians offer figures ranging from the low millions to a high of 200 million. An accurate count we'll probably never have, though the Tatanka Center in South Dakota offers 30 million as the most likely number based on the carrying capacity of the Great Plains.[5] One thing is abundantly clear. In the American West, the millions of living, breathing, grunting, snorting, mate-guarding, and playing bison were killed off one by one by one, until they numbered in the mere hundreds.[6]

The image of rich men and women handing over a fistful of dollars for the privilege of chartering a train car, then riding out into the open and killing buffalo, is as vivid as it is sickening: "The well-heeled simply removed their top hats and shawls, leaned out the windows, and fired."[7] Yet it's all too easy, and wrong, to demonize any single group. Long before the railroad tracked across the West, buffalo had

been shot and killed in significant numbers by Oregon Trail pioneers
and by buffalo hunters both Anglo and Indian. Certainly the U.S. government cannot escape implication. The Tatanka Center puts one perspective forward quite bluntly: "[W]hen Euro-Americans realized that
the native culture would change drastically if buffalo were eliminated,
it seemed like a concerted effort to eradicate the animal to change the
people. In the process of destroying the buffalo population, hunters
cleared the way for Westward expansion and the old way of life of the
Plains Indians disappeared."[8]

The human-buffalo relationship is ancient. As we know, ancestors
of today's buffalo appeared in the glorious cave art of ancient *Homo
sapiens.* As Ken Zontek notes in his book *Buffalo Nation,* the archaeology of the region around today's Yellowstone National Park shows
that by about 10,000 years ago, the Indian-buffalo connection was
firmly established there—so close to where I sat in our car, immersed
in the herd surrounding us. And for some Indian groups on the Plains,
like the Lakota Sioux, the bison (called *tatanka* in the Lakota language)
was the center of all life.

Part of the Great Sioux Nation, the Lakota people number about
70,000 today. In traditional times, the Lakota centered their settlements in North and South Dakota. Often thought of as nomadic
hunters and gatherers, famous for hunting buffalo on the Great Plains,
the Lakota were in fact primarily agriculturalists until the eighteenth
century. They farmed intensively, until they acquired the domesticated
horse, an event that changed their culture markedly—and their relationship with the buffalo along with it. Hunting was now possible in a
new way. The Lakota and the buffalo merged so that the people's lives
cannot be understood without taking into account the buffalo, and
the buffaloes' lives cannot be understood without taking into account
the people.

When a buffalo was killed by the Lakota, the meat was consumed
and the hide, bones, and horns were used for clothing and shelter.
Every last scrap of the buffalo was employed to some productive purpose. The bladder became a water container, and the dung was put to

use as fuel. For the Lakota, as we will see, a relationship with the buffalo was a gift, and to waste even a small part of that gift was to dishonor the Earth.

It comes as no surprise, then, that as the buffalo declined, the very fabric of Lakota lives altered drastically. No loss exceeded that of the sacred.

The Lakota tell the story of the White Buffalo Cow Woman, who brought gifts to the Lakota people when she appeared on Earth. These gifts were sacred, to be used in ritual. Among them was a redstone pipe with a carved buffalo head on the bowl. The White Buffalo Cow Woman explained that when a Lakota person smokes a sacred pipe, Wakan Tanka, the Great Spirit, may hear his voice.

Thus was forged a sacred covenant between the buffalo and the Lakota nations.[9] In the book *The Sacred Pipe,* Black Elk recounts the events involving the White Buffalo Cow Woman and their significance.[10] It is important to realize the authority with which Black Elk speaks. Born in 1863, he represents a touchstone figure for many critical, even tragic, events in the lives of the Lakota. In 1876, when he was not yet an adult, Black Elk witnessed the Battle of the Little Bighorn. At one point in his life he became part of Buffalo Bill's Wild West Show, and later was injured at the famous Wounded Knee Massacre in 1890. He is most remembered, and indeed revered, for his authority on Lakota spiritual matters.

Black Elk explains that as she finished her visit to Earth, the White Buffalo Cow Woman said this: "Behold this pipe! Always remember how sacred it is, and treat it as such, for it will take you to the end. Remember, in me there are four ages. I am leaving now, but I shall look back upon your people in every age, and at the end I shall return."[11]

With that, she walked off, but soon looked back and sat down:

When she rose the people were amazed to see that she had become a young red and brown buffalo calf. Then this calf walked farther, lay down, and rolled, looking back at the people, and when she got up she was a white buffalo. Again the white buffalo walked farther and rolled on the ground, becoming now a black buffalo. This buffalo then walked farther away from the people, stopped, and after bowing to each of the four quarters of the universe, disappeared over the hill.[12]

The first remarkable thing about this passage is its demonstration of the intense bond between the Lakota and the buffalo. Through Black Elk's recounting we learn that, for the Lakota, the buffalo were "the closest living relative" among all four-legged creatures. What an amazing phrase! Today we know that chimpanzees and bonobos are humans' closest living relatives, and routinely use that phrase to convey, in a scientific context, the significance of the human-ape relationship. In the Lakota case, the phrase conveys a sacred relationship.

Second, it is not just a sacred closeness in word and idea: rather, the closeness is embodied. Note the physicality conveyed in the passage above: the sacred woman *becomes* a buffalo. This transformation may bring to mind the focus on animal-human fluidity that marks some of the most arresting images on ancient cave walls, and also the religiosity found among the forest-dwelling Runa people of Ecuador and the reindeer peoples of Siberia.

Lastly, the white buffalo, one of the forms assumed by the sacred woman that day, became a sacred symbol for the Lakota, and remains so. The birth of a white buffalo calf is met with enormous excitement. In summer 2007, for instance, a white buffalo calf was born to White Cloud, a rare albino buffalo housed at the National Buffalo Museum in North Dakota. White Cloud's four previous calves were brown, but the new birth appeared pure white. The birth generated enormous interest, some of it owing to the sacred nature of the color.[13]

Other tribes than the Lakota also link their identity to that of the bison. The Crow, Cree, and Arapaho nations embrace creation stories that point precisely to the bison. Sometimes the native perspective is explicitly evolutionary. A man named C. Wolf Smoke gave a lecture in

Nebraska in 2000 and, according to Ken Zontek, said on behalf of other Indians, "We evolved from the bison, we used to be bison. If you accept Darwin, then you should accept this."[14] The Lakota make a good case study, but we should not lose sight of the fact that for a number of Indian groups, loss of the bison meant something very much like a rift from their own origins.

The human-buffalo history is complex, sticky to untangle, and varied in its emotional tenor. Anglo and Indian communities, in their different ways, revered the bison *and* contributed to its mass slaughter. For our species, cruelty and compassion combine, time and time again—all too often, to the detriment of the animals in our world. The bison's fate remains uncertain even now.

In Wyoming and Montana, the fate of the bison is at best a mixed picture, despite on-the-ground activism to protect buffalo, by the Buffalo Field Campaign (BFC) and other grassroots organizations. The idyllic experience I had in Wyoming, observing close at hand a protected herd, too often has no parallel in neighboring Montana. In the winter of 2005–6, 970 Yellowstone bison were killed by state and federal agents. This number was uncomfortably close to the toll from winter 1996–97, when nearly one thousand bison were felled by bullets.[15] In a tragic turn of events during winter 2007–8, that previous record was broken when over 1,500 bison were killed.

This slaughter goes on even though the Yellowstone herd is close to unique: these are pure bison, who share no genes with cattle. Why does this happen?

The first answer relates to ecology. Wyoming's Yellowstone is at high elevations, with a great deal of snow cover; in winter, the buffalo move to lower elevations in order to find enough food to survive. This means they cross into Montana. And there they become nothing so much as disease carriers, in the state government's eyes. The Department of Livestock in Montana views the bison as vectors for brucellosis. The BFC explains: "Brucellosis is a disease that can cause spontaneous abortions in cattle. . . . There has, however, never been a confirmed case of brucellosis transmission from buffalo to cattle under natural conditions. Indeed, in Grand Teton National Park, where infected buffalo and livestock have commingled for more than forty-five years, there has not been a single incident of disease transmission."[16]

The BFC is an activist organization. BFC volunteers station themselves out in the cold, in a wintry and snowy national park, on the very front lines of this battle; there they disrupt and document Montana's practices of harassment and killing of the bison. (Consult the BFC video footage at www.buffalofieldcampaign.org.) This group does not aim for an objective stance. Yet it is clear to Roger Brooks, the companion and caretaker of Charlie the buffalo, who remains unaffiliated with the BFC, that Montana's response to potential brucellosis contamination of cattle by bison is a serious and prolonged overreaction.

Brooks's view is not black-and-white. He supports the BFC's campaign to protect the Yellowstone bison, but worries that the activists are politically naïve in pushing for solutions that completely circumvent the ranchers' worries about brucellosis. "Caught between intractable idealists and cattle-culture fundamentalists," Brooks feels the best solution may be to work toward two practical measures: a bison-specific brucellosis vaccine, and a highly accurate no-kill method for testing bison for brucellosis.[17]

Even as the struggle in Montana continues, Anglos and Indians are joining together in their aims for the bison. In the early 1990s, prominent leaders in the Indian bison-restoration movement held a ceremony to ask the bison themselves if they wished to return to their lands: "The leaders received an affirmative answer, validating their efforts past, present, and future."[18] The InterTribal Bison Cooperative (ITBC) was formed, with a collective understanding that "reintroduction of the buffalo to tribal lands will help heal the spirit of both the Indian people and the buffalo." As of late 2007, fifty-seven tribes belonged to the collective and 15,000 bison were managed under the group's aegis.[19]

Consider the fact that it would be possible for a meeting of the BFC or the ITBC to occur in Yellowstone—in Wyoming's Yellowstone, mind you, where buffalo are protected—and finish up with a lunch at which bison sausages were served. Indeed, the ITBC offers buffalo-based recipes on its website![20]

BISON, CATTLE, AND THE FARMING INDUSTRY

The bison is simultaneously a key part of ecosystems that draw millions of people toward wildlife-viewing, an integral part of the religious worldview of native groups, and meat for our stomachs. One day when I was caught up in reading about the spiritual nature of the bison in Native America, what did I find in my mailbox but an envelope with a picture of grilled meat on the front, and these words: "Bison: Better for You than Beef." On the envelope's back was a chart listing the nutritional benefits of grass-fed bison compared to beef, salmon, and chicken, and a come-on to find "more juicy details inside." There I found a small piece of paper that said: "The most-often-asked question about bison is. . . . How long should I cook it?"

I don't know if that's the most-often-asked question about bison. I hope it is not.

Yet there's no escaping the fact that buffalo is hot right now in the fine-dining market. One Wednesday, the *New York Times*'s dining section boasted a large photograph of a bison on a South Dakota ranch, paired with the headline "Home Again on the Kitchen Range." According to the article, buffalo ranching took off in the 1990s, only to meet with consumer resistance. People wanted the familiar, not what they perceived as "tough and gamey" meat. Later, in the mid-2000s, things changed: "Today buffalo meat, shunned no longer, has achieved an enviable position: simultaneous praise from chefs, nutritionists and environmentalists. At last, steak without guilt."

Rancher Mimi Hillenbrand has this to say about buffalo: "You can't help but love them: they are smart, intelligent. When you are out with them every day it's magical. People who raise buffalo absolutely love their animals, unlike those who raise beef. For them it's all about price. For us it's all about restoring the prairie." For a moment, it's easy to forget that her comment appears in an article about raising buffalo for *food*.[21]

Hillenbrand, it seems, feels no contradiction between the magic of loving buffalo and the magic of eating them. I could not move so fluidly and guiltlessly between these two states. But that's not because I am driven by strict principles of vegetarianism. True, I eat no beef, pork, veal, or lamb, as well as no bison. I do eat some chicken and fish. Who am I to set out principles of what's okay and what's not okay for others to eat?

Let's dig deeper. Can eating animals be part and parcel of a humane approach to animals? How an animal is brought to the table for human consumption matters a great deal. Turning buffalo into food on a ranch is not the same process as killing bison in Montana as they wander outside the boundaries of Yellowstone National Park. To live as a grass-fed buffalo in the open on a ranch is not to live as buffalo sometimes are forced to live in Montana. According to an update from the BFC:

> Two of the buffalo that were killed near West Yellowstone yesterday, right next to Highway 191 along the Madison River, were females with calves. Patrols witnessed the calves this morning, acting very distressed. They were not with the nearby herd, but instead with the remains of their mothers. They were acting frantically, making heart-wrenching bellowing sounds, running back and forth and coming back to the kill sites. They must feel so incredibly lost right now. These little calves are just six months old and still nursing.[22]

To die is to die is to die? Or are there better and gentler deaths for animals, just as there are for humans? In her books *Animals Make Us Human* and *Animals in Translation,* Temple Grandin stressed the importance of working not just toward societal ideals like vegetarianism, but also toward practical goals such as humane slaughter. When farm animals like cattle and pigs die, they need not die in terrifying and painful ways. Could the same principle be applied to buffalo?

Native American groups are heavily involved in the buffalo-as-commercial-product game. Making its debut in 2007, for example, was the Tanka Bar. On its bright wrapper, the image of a running person is superimposed on that of a bison.

Created by the Oglala Sioux on the Pine Ridge reservation, the Tanka Bar is made of buffalo meat and cranberries. Its marketers explicitly link the snack's goodness with traditional Indian lifeways: "We're convinced our Ancestors knew what they were doing. In times when Native Americans sustained themselves with wasna [a mix of meat and berries] and other traditional foods, heart disease, diabetes, or obesity were virtually unknown. Tanka products honor their wisdom . . . the Tanka Bar offers powerful protein for your life journey in a bar so good, you can Taste the Energy!"[23]

We find here an eagerness for turning bison, a symbol of our country's history and a religious touchstone for some in a way that the cow or pig has never been, into food. Is this a modern version of traditional tribal waste-nothing practices? Is it just hard-core realism, given that buffalo ranching is with us to stay? Or is this crass exploitation of an animal with a supposedly special status in the United States? Might it be all of these things at once?

None of this cultural schizophrenia about bison should surprise us too much. How many of us were raised on cute animal stories that featured talking—and wise—cows, pigs, and chickens (or frogs and toads), even as the dinner served to us that evening boasted beef, pork, or poultry? Fantasy figures, partners in fulfilling relationships, meat for the table—animals have held multiple and overlapping roles in humans' perceptions since the Ice Age.

In sum, bison are important to the religious imagination of

thousands of people. For many thousands more, the buffalo connects people to their land, and to a world of majestic animal beauty. Yet they suffer at our hands, and when animals suffer because of us, we deny ourselves a part of our humanity. And we may deny ourselves good health, too.

BETTER LIVING THROUGH ANIMALS

Those of us who invite animals into our lives reap direct physical and emotional benefits. Consider these findings:

- Quiet stroking and petting of a dog can, within half an hour, reduce a person's blood pressure by about 10 percent. Intriguingly, the blood pressure of the *dog* drops immediately when the dog is petted. What a great example of the mutual effects of animal-human relating, felt in the body as well as the spirit!
- In at least one study, heart attack patients who own dogs are significantly more likely to be alive a year later than those without dogs. Whether this is because of some kind of emotional high from closeness to an affectionate animal, or merely because of the healthful benefits of walking a dog, isn't entirely clear. Both factors may work in tandem.
- Swedish scientists showed that children suffered from fewer allergies and less asthma if they lived with pets during their first year. Pet-averse family and friends sometimes admonish new parents that babies and pets don't mix—but they may, if basic safety precautions are followed.
- Cutting-edge research suggests that dogs may be able to detect a wide variety of cancers via their exquisitely attuned sense of smell. One study shows what happens when dogs are trained (with rewards) to discriminate between two types of odors:

those collected from people with diagnosed lung and breast cancers, and those collected from healthy people. When these dogs are asked to sniff out cancers from a new group of people, some healthy and some ill, they perform at highly accurate levels. The accuracy holds even when cancers are at an early stage. Someday, dogs may aid in cancer diagnosis, as a living tool in the doctor's arsenal. One case study shows that a dog alerted his owner to melanoma by constantly sniffing a lesion on the person's skin.[24]

These results range from the logical (petting an animal for a long period can be calming) to the astonishing (dogs may detect a serious illness). The positive aspects of animal-human relating are of keen interest to scientists these days, and there's more than enough room in applied health care for visionary ideas based on animal therapy. In 1997, registered nurse Jennifer Jorgenson wrote an article for the *Journal of Nursing Scholarship* in which she asked: "Could patient motivation to get outside and throw a ball or go for a walk be increased when patients have dogs? Could we [nurses] further decrease stress by providing a focus away from a medical procedure such as drawing blood or changing a dressing? Would the presence of a dog create a more balanced relationship between a nurse and patient to foster an equal partnership in care?" Jorgenson concludes, "Nurses are in a perfect position to pursue Animal Assisted Therapy research. Our philosophy embraces total well-being. Animal Assisted Therapy combines both the physical and emotional components of health."[25] Jorgenson's questions, still timely, point us in promising directions for the future easing of patients' lives.

Exploring the link between animal-human relating and our emotional health is especially instructive when we take a look at children. Picture thirty boys between the ages of twelve and fifteen, all diagnosed with a type of behavioral disorder that involves an inability to inhibit aggression. When they get upset, the boys tend to "get physical," to push or grab or hit, so much so that they are living in a residential treatment school.[26] They tend to ignore rules, and fail to respond to rewards or to punishments.

Half the boys are assigned to an Outward Bound program for six months, and the other half to what's called the Companionable Zoo. Developed by Dr. Aaron Katcher, the Companionable Zoo is a collection, maintained on school grounds, of over a hundred small animals, like rabbits, guinea pigs, iguanas, turtles, and finches. An occasional larger mammal, such as a goat or potbellied pig, can be found as well. *Why the Wild Things Are,* written by Gail Melson, includes the story of what happened when the boys and the animals were brought together.

For five hours each week, the boys took animal-care classes. Once they had acquired a certain level of knowledge about an animal species, they could adopt a pet and involve themselves in its life, even taking it on field trips to hospitals and special-education classes. "Two cardinal rules reigned at the Zoo," writes Melson, "speak softly and be gentle with the animals, and respect the animals and each other."[27] The behavioral results were swift and, to those at the school, unexpected. None of the boys were cruel or even careless with the animals; they readily followed rules for animal care. Most dramatic were the nurturing, cooperative, affectionate, responsible, and playful behaviors shown at the Zoo by the boys.

The Outward Bound program represented a kind of scientific control because it involved no sustained animal contact. The boys in that program continued to exhibit the same high levels of aggression and need for physical restraint as they had upon entry.

When the six months ended, the two groups switched: that is, the boys in the Outward Bound group became immersed in the Zoo, and vice versa. Unfortunately, some of the "Zoo boys" began to act out again. The take-home lesson is not that the Zoo program failed to imbue boys with any lasting benefits. Rather, the lesson is one of individual variability. The boys who, during the Zoo program, had been rated by their teachers as engaged, motivated, gentle, and respectful, declined steadily in disruptive behavior, including classroom aggression. Six months after the program ended altogether, the high-performing Zoo boys continued to do well. Katcher's Zoo program has been extended to include children with autism and similar developmental difficulties.

In her book *Made for Each Other*, science journalist Meg Olmert also describes Katcher's work with children and animals. For Olmert, though, the transformation witnessed in the boys is largely due to oxytocin, a hormone that cascades through the body during person-person and animal-person interactions. For Olmert, oxytocin is the power behind attachment, attunement, and the other processes I have been describing. She even attributes the animal-domestication process to the bodily "chemistry [that] seems to have gained a grip on all humanity and the animals they embraced."[28]

It's worth pointing out that Olmert's oxytocin story speaks to the grip on our society of all things biological (DNA, oxytocin, and so on). To be sure, there's a biological basis to animal-human relating. Oxytocin may produce short-term changes in mood and an openness to experience. But it's poor anthropological practice to award more power to oxytocin than to people's individual agency. While oxytocin may give someone a window of opportunity, a biological moment of possibility for enhanced relating with animals, what happens next is up to that person's feelings, thoughts, and actions. (For details on this debate, see my review of Olmert's book.[29])

Returning to Katcher's troubled boys, I see them as a dramatic example of Melson's theme. Melson elaborates on four ways in which a child's life may be transformed by relating with an animal; they apply to any child's life, not just to those with a behavioral disorder. I read and reflected upon these as a scientist, but also as a mother.

My husband and I laugh about it now, but when our daughter was three, we all seemed to live life on the edge of intensity. Sarah's single mode of locomotion at that stage was the headlong run! Wherever she wanted to go, in the house or outside, in a safe backyard or on a city street, she'd put her head down and charge ahead. Everything in her life—and as a result, in ours, too—seemed to unfold at top speed! With this intensity came frustration in the form of clenched fists or tears when Sarah's motor skills weren't equal to the goals she sought. When she struggled to feed herself, or to master challenges like small buttons or wayward shoelaces, Sarah, like most kids, sometimes lost control of her emotions.

For a small child, life is full of joyful adventures and small set-backs. To keep on an even keel isn't always easy. It's a skill to be learned, right along with wielding a fork or lacing up shoes. In Sarah's life there was a single creature even more dependent, in some ways at least, than she. Sarah grew up with Swirl, a gentle gray-and-white swirl-patterned cat, and having Swirl nearby helped Sarah moderate her moods. Nearby was a soft creature ready to be petted—indeed, who had to be treated gently every single time, or Mom and Dad would intervene. As Melson puts it, "animals' combination of arous-ing appeal and dependence may help young children as they struggle to command their emotional swings."[30]

Being with a nonverbal (meowing!) creature may induce in a child a watchfulness, an ability to pick up on body-language cues, and a growing recognition that we humans have a responsibility beyond just ourselves. Swirl was a patient saint among cats, when it came to being with a young child. She never hissed or clawed, and seemed to forgive Sarah's well-intentioned attempts to pick her up. Even Swirl had her limits, though. Sarah came to learn when Swirl had had enough attention and wanted to be on her own; all cats are marvelous teachers in this regard, and Swirl had a particularly expressive face. When she felt disdain, her face reflected her emotion perfectly, and some human or other would become the recipient of her haughty stare.

Melson points out that child-pet communication "may help chil-dren hone their skills in picking up cues to internal states of other beings from body movements, facial expressions, and voice tone or pitch, all important components of emotional intelligence."[31] Here we see the developmental basis for empathy and compassion.

Wouldn't Sarah have learned just as much—indeed, more—from a younger brother or sister as from her pet cat? Like Melson, I think the answer here is likely to be no. Swirl did for Sarah what no sibling could possibly have done: day after day, week after week—indeed, year after year—she offered a companionable presence without any teasing or tense rivalry or any of the other complex components of sibling relationships.

There's no doubt about it—relating with pets can at times involve challenges and frustrations. But in almost all instances, these pale next to the wonders, as Melson writes: "Children feel a sense of unconditional love from their pets. . . . Dialogues with pets offer a time-out from the anxieties of human interchange." And it's a time-out that goes beyond the physical release of playing in the backyard, or watching a favorite TV show, because "only a pet provides a sentient, feeling presence."[32] Here, I think, many adults will nod in recognition. While the unconditional-love angle is significant developmentally, it's an aspect of animal-human relating, at least pet-human relating, that doesn't change with a person's age.

Finally, in tandem with Melson's fourth point, I can't resist bringing up the *Frog and Toad* books once more. From Sarah's earliest weeks of life, we read to her the adventures of this amphibian pair. As we have seen, life lessons were conveyed by Lobel in the lightest of manners: no heavy moralizing, but points about loyalty and shared joys are communicated. Frog and Toad hold a special place in my memory, but a look at any bookstore shelf or book catalog with a good children's section will tell you that they're representative of a pattern. "It is remarkable," notes Melson, "that animals loom as such important characters, not only in the literature and art produced *for* children by adults, but also by children in their play, dreams and stories . . . animal stories and symbols guide children into deeper understanding of what it means to have a *human* self."[33]

In sum, the peculiar double-edged awareness that we humans are at once *of* animals and *for* animals, that we are part of the world's vital web but uniquely able among all species to care for it, actually helps us care for ourselves. Our humanity renews itself and expands because working for animals and working for people tap into the same impulses for good action in the world.

In his book *The World Without Us*, Alan Weisman imagines what the world would be like without humans. In Manhattan, within two days the subways would flood (because no humans could pump them out). Within twenty years the Americas would reunite (because the Panama Canal would have closed). Within 100 years, the dwindling elephant population would increase twentyfold (because no human poachers would kill elephants for ivory or meat).[34]

What I imagine is a world in which an ethic of kindness and compassion for all creatures—rooted in God, or rooted in beyond-the-self spirituality, or rooted in an intense secular love for the natural world—informs all of human behavior. In this world, Miss Oinkers is loved by her family, even though a warm outdoor enclosure had to be built to accommodate her. In this world, animals are spayed and neutered as a matter of course, so that great numbers of unwanted and hungry feral cats become a thing of the past. In this world, the bison roam free and unharmed in Montana.

When we relate positively with animals—in urban and suburban settings as much as in any other—all parties reap the benefits. That animal-human relating is not always positive for the animals involved is an inescapable reality. At times the complexities of our past patterns of relating with animals—the side-by-side spearing them and spiritualizing them—may manifest today in terrible habits of neglect and cruelty, patterns that exist right alongside the sometimes excessive pampering of our pets.

But for many thousands of years now, our evolutionary history has been in large part directed by cultural practices. We shape our future now with more power and precision than when we became predator animals as well as prey, or when we domesticated the dog, the sheep, and the horse. We have the power to nurture in our children the skills of empathy and compassion that have evolved in us to a degree unprecedented within the animal kingdom. We owe it to our evolutionary partners, the animals, to do this.

11

Clones, Crows, and Our Future

• • • •

> The parrot . . . had flown to the highest branches of the
> mango tree when they took him from his cage to clip his
> wings. He was a deplumed, maniacal parrot who did not
> speak when asked to but only when it was least expected,
> but then he did so with a clarity and rationality that were
> uncommon among human beings.

—From *Love in the Time of Cholera* by Gabriel Garcia Márquez

> When people laugh at Mickey Mouse, it's because he's so
> human; and that is the secret of his popularity.

—Walt Disney

W E S E E O U R S E L V E S in other animals—in their everyday
expected behaviors and their rare unexpected behaviors.
Being with animals brings forth the deep knowledge that we once
were animals, and are animals still.

We crave new ways to be with animals and work to create them.
Browsing the online catalog for Elderhostel, the program for lifelong
learners age fifty-five and up, I was startled to come across a course
called "Intergenerational Mammoth Hunting Under the Midnight

Sun in Alaska." Designed for grandparents to share with their grand-children, the program promises an anthropological adventure among the Inupiat Eskimo peoples, hunters who live north of the Arctic Circle, on the edge of the Chukchi Sea.

But wait a moment, *mammoth* hunting? Mammoths, the shaggy, curvy-tusked beasts of the Ice Age, went extinct thousands of years ago. On close reading, it became clear that program participants search the area for mammoth *bones*. That's exciting enough, an experience designed to put people in touch with the ancient past, when mammoths walked the Earth along with our own ancestors. Wait a decade or two, though, and it may be possible for Elderhostel to offer a close encounter of a different kind.

In late 2008, scientists announced the decoding of the mammoth genome.[1] The human genome (or at least the genome of a few human individuals) had already been sequenced, as had that of some other mammals. The mammoth case is different because that species has been long extinct, a fact that raises the specter of cloning—and resurrection of ancient life. As the science writer Henry Nicholls puts it, "If you want to bring a species back to life, the mammoth would be almost as dramatic as a dinosaur."[2] True enough: with its enormous bulk and ability to evoke Ice Age hunting cultures, the mammoth's arrival into the twenty-first century would feel to us like a peculiar form of time travel.

Nicholls is quick to point out that we don't yet have the technology to bring a mammoth back. Scientists must work from more than a single genome and ensure a ready surrogate home for a cloned fetus. Years upon years and dollars upon dollars would be required to complete the project, even once these initial challenges were met. Still, it was only fifteen years ago that almost no one believed a mammal could be cloned at all; Nicholls rules out nothing.

I have written much in these pages about how our present being with animals emerges from our past. We humans are symbolic thinkers, emotionally resonant with the natural world, because we have coevolved with animals from our scavenger-hunter-gatherer days, through Ice Age rituals and Neolithic burial rites to the ongoing, ever-dynamic processes involved in domesticating other species. The earliest temples and monumental sculptures reflected our ancient linking of the supernatural with animals, and, more than that, our *engaging* with the supernatural through animals.

But what about the future? Humans evolved, and we evolve still. Even now, with the powerful forces of culture that I have emphasized time and time again, we are still sculpted to an extent by biology. We are, through and through, the *biocultural* species. Our backs readily pain us because human bodies are an anatomical compromise between efficient bipedalism and the need to bear big-brained babies. Human infants are born neurologically incomplete, vulnerable, and in need of round-the-clock care, because a "finished" brain cannot fit through a woman's birth canal without severely endangering her life. As a result, human infants at birth are so neurologically underdeveloped that they must learn how to breathe during their first few months.[3]

Evolution is all about adaptation to the environment, but many modern humans are no longer so well adapted to theirs. Too many of us overindulge in the sugar, fat, and salt that our hominid ancestors could never get enough of. A joke in bioanthropological circles is that the next human species to evolve, the one to replace *Homo sapiens,* will be *Homo sedentarius obesus.* Big-brained and computer-savvy, they will be physically unfit to the max. That joke isn't really so funny, because it indicates a sorry mismatch between our biology and our culture, one that manifests in the current obesity epidemic.

Fortunately, the situation with animal-human relating is different. That animals matter to us so deeply is in part a product of our evolutionary past, but how we envision and create modern cross-species relationships is not constrained by our past. To hunt other animals and to eat them are not hard-wired human behaviors, as can be seen in the growing vegetarian and vegan movements. Also, the very thesis of this book is that humans evolved to observe other animals so keenly as to comprehend the malleability of their individual lives; to bring those individual lives into our communities and our homes; to usher animals into our religious lives in the daily present and the otherworldly future; and above all to feel animals' sufferings and joys—and, because we do, to feel more deeply our own.

So what *about* the future of animal-human relating? Maybe a cloned mammoth won't grace our immediate future. I hope not. As one editorial writer put it, "The first mammoth would be a lonely zoo freak, vulnerable to diseases unknown to its ancestors. To live a full and rewarding life, it would need other mammoths to hang out with, a mate to produce a family and a suitable place to live. The sort of environment it is used to—the frigid wastes of Siberia and North America—are disappearing all too fast."[4]

Society has already ventured partway into this territory. The roster of living mammals that already have been cloned includes mice, water buffalo, and even a rhesus monkey. Dolly the sheep became famous, headlined around the world as the first cloned mammal, though technically she was not a clone at all. (A clone is by definition a copy of a single parent; Dolly had two parents and was produced via a slightly different technique.) If not a clone, Dolly most assuredly was a symbol of humans' genetic tampering with life, perhaps even of our playing God, indeed of a whole imagined future of altered life feared by some and welcomed by others.[5]

How would our relating with animals change if cloning became a frequent practice? To think about this question involves thinking about the motivations for cloning. At least three different goals exist. Cloned animals might amplify and enhance the world's food supply. In 2007 the U.S. Food and Drug Administration gave preliminary approval for the use of cloned animals to make food. Best-practice guidelines are available for how to judge, for instance, the "steakworthiness" of dead steer: how to figure out when it's worth extracting from a side of beef the cells that can become a calf that in turn can become someone's dinner.[6] If this scenario is repellent to some, to others it is business as usual. An Australian champion of cloned foods notes dryly that humans have been eating and drinking clones for centuries: every glass of wine we enjoy, for example, involves the ingestion of cloned grapes.[7] To mix metaphors, whether to compare cloned calves with cloned grapes amounts to comparing apples and oranges remains in the eye of the beholder. Complicating the situation is the fact that conventional raising of animals for food is hardly a benign process; see Jim Mason's and Mary Finelli's writing on factory farms.[8]

A topic that's so vast as to exceed the bounds of reasonable discussion here is the animal cloning that is undertaken in order to propel medical advances. Small mammals such as rabbits are considered by some as good models for studying the diseases of the human heart and blood vessels; cloned pigs and monkeys are close analogs, in certain ways, to human bodies. The ethical debates are even fiercer for this type of biomedicine than for cloned foods: Do human needs inevitably trump those of animals?

A human need, right alongside that for food and good health, is emotional companionship with pets, and a third justification for cloning emerges directly from it: what I call the pet-replicant trade. (Replicant: Think of the movie *Blade Runner*, starring Harrison Ford.) A company in South Korea is one among a few to offer a dog-cloning service. The cost is $150,000; because there's only a 25 percent chance of success in any given case, the client pays once the replicant pet is in his hands. A salesman for this outfit commented, "Canines die faster than humans, but now people can have the same dog for their whole lives."[9]

This statement gives me pause. Does ordering up a replacement for a cherished dog that has died separate us too completely from a natural cycle of birth, life, and death? What about the claim that a clone is *the same animal* as the original beloved pet? Any creature's personality emerges from a dynamic mix of genetic influences and life experiences. No geneticist could be surprised by a report in the *New York Times* that some of the South Korean dog replicants vary substantially in appearance and behavior from their parent "originals"— or that some clients feel less than thrilled, as a result, with their high-priced product.[10] Even more troubling are the costs to the animals themselves: the Humane Society of the United States condemns pet cloning in large part because of the high rates of trauma and early mortality involved.[11]

To the range of ethical challenges already considered in this book, then—those involving Africa's chimpanzees and gorillas killed for beef; Yellowstone's bison shot when they wander from Wyoming to Montana; and the homeless pets and feral cats and dogs that populate all our cities and towns—can be added those that involve DNA manipulation. Christian theologian Andrew Linzey pulls no punches on the topic. The genetic tampering with animals "so that they become only means to human ends" is, Linzey says, "morally equivalent to the institutionalization of human slavery."[12]

With books like *Animal Theology*, we turn from science to religion. Linzey stands at the forefront of a wave of change sweeping through Christian theology in its stance toward animals. And Linzey is far from alone. Religious thinkers in Islam, Judaism, Hinduism, and other faiths, some of whom I have described in these pages, speak out for animal welfare and gently move their congregations toward reverence for animals. Though not new, this movement is gaining new momentum in a world where *Homo sapiens*, the Dominant Animal, causes unprecedented changes to the natural world.[13] Religion and animals seem a natural fit; they've been twinned since the dawn of our species, and the emotions engendered by that twinning are often palpable today. I will never forget the communal joy that lighted the very air inside the Cathedral of St. John the Divine on Saint Francis's Day, when I watched Sally the camel lead a glorious animal procession through the sacred doors, and when I shared animal lovers' delight at receiving for their dogs, cats, rabbits, and birds a personal blessing from the clergy.

Reverence for animals need not occur inside the walls of a cathedral, mosque, synagogue, or other house of worship. This book reveals the anthropology of why we are obsessed with animals, whether at worship or at home, whether out in our community or hiking in a national park. I have also written some about how my academic life, with its focus on how an ape evolutionary platform led to human culture, language, and religion, meshes with my at-home life, so graced by the presence of rescued animals. It is fitting, then, that I make my last animal story in this book one that touches on the personal and the everyday, but also on science, and indeed on something more. The story centers around my friend Nuala Galbari and an American crow named Reginald. Nuala and Reginald, or Reggie as he is nicknamed, have been together for sixteen years. Theirs is a cross-species friendship in the mold of real-life Owen and Mzee, or fictional Frog and Toad. In a sense, it's a story about all of us, our bonds with the animals in our lives and the potential for deepening those bonds.

Nuala lives with Reggie (and her human partner, and a number of rescued cats) in Virginia, where he occupies a spacious outdoor aviary. Reggie is crippled, but otherwise healthy and able to perch. On several visits to their home, it was clear to me that Reggie has adapted very well to captivity and to the companionship of his human caretakers. He also enjoys chances to interact (by sight and sound) with the wild birds of the neighborhood. Reggie sometimes vocalizes in order to "call in" his wild counterparts, who then approach his aviary. He is responsive to Nuala's presence, and from all that I can tell, he trusts her, his closest friend.

Nuala and Reggie moved to Virginia from the much colder state of Minnesota. Virginia nights are warm or at least mild for a good portion of the year. From December through February, however, the temperatures may plunge; Nuala often brought Reggie indoors during these months, despite the working heater in his quarters, in order to ensure his protection from the cold. However, in early winter 2008, she tried something new. No longer would Nuala assume that she knew when Reggie wished to come indoors in the evenings; instead she would invite him to tell her.[14] "Each night," says Nuala, "instead of stressing the poor bird by gently picking him up (to which action he often complained), I thought I would quietly ask him, 'Would you like to come in tonight?' Now, I know this may not be crow language exactly, but I felt that he would understand me if I used *the same words* and *the same intonation* each time. He would know (or would learn) what I meant."

From that point on, the woman and the bird participated in a process of mutual attunement and learning. Without using any food treats or other rewards, Nuala queried Reggie with the same question each night at dusk, while showing him his carrier cage: Did he want to come indoors?

A writer, Nuala tells what happened next in her own words:

The first night, Reggie looked at me as though he didn't know what I was trying to do. He didn't respond, so I left him out.

The second night, he moved around as though trying to work out what it was I wanted. He was a little nervous, but remained on the perch. I left him out.

The third night, Panther, the black cat, was lurking around, and Reggie appeared jumpy. I told him it was all right. He remained in the corner of the aviary.

As I repeated this request each night, I noticed he was engaged in the process and seemed to be figuring it out.

By the fifth night, I opened the aviary door, and set the carrier just at the entrance. I said, quietly, "Would you like to come in tonight?" He fell off the perch, and then walked over to the carrier and stepped in. He slept quietly, in the warmth of the living room, with a view to the deck. In the morning, he hammered on the sides of the carrier to let me know he wanted to get out. I then talked to him and took him back out to the aviary.

The process was repeated for two more days and I noticed a faster response on each occasion. . . . Now he jumps into the carrier within thirty seconds, the moment I ask. I know this is not from any fear, as I have not forced him to do one thing or the other—merely asked and on each occasion, he has chosen.

It's gotten to the point now where Nuala's partner, David Justis, can approach the aviary and stand quietly, with the bird carrier open. With no verbal prompting from David, Reggie immediately jumps into the carrier. Importantly, though, this response is not automatically triggered by the sight of Nuala, David, or the carrier: Reggie opts whether to come inside or not, apparently depending on the degree of cold or darkness at the time of the offer. Some nights, he decides to stay out. And if Nuala becomes slightly impatient and raises her voice, Reggie's movements become hurried and nervous, and his pupils dilate; when she returns to her normal, calm voice, he relaxes.

And sometimes Reggie chooses to surprise! One night he acted in a different way: "The carrier was not set down, just held in midair. Reggie walked sideways on the aviary perch toward his dish, picked up two grapes, and then jumped into the carrier. It was as though he was now being a comedian, and creative to boot."

Nuala's desire to attend to Reggie's preferences, and to trust that she and he together could communicate across species lines about something entirely novel, reflects a high degree of attunement. We know that corvid species—American crows, ravens, and the like—are smart birds: that's a scientific fact. (Remember Craig Childs's encounter in Utah with the owl-feather-caching corvids?) Yet I think it matters little for us to try to parse whether Reggie comprehended Nuala's actual words, or her voice tone and gestures. The story speaks to more than bird learning or cognition, and that "more" comes across as Nuala reflects upon the experience with Reggie:

> This has become a regular situation, now. It is, to me, a sign that Reginald and I are a team and that he knows I will look after him. More than that, I think he knows I am attempting to communicate with him in a stronger, more patient way, rather than deciding what I want for him. . . . When I stroked his feathers today, he placed a claw gently on my hand. Silken bird.

In Nuala's account, I see a way forward for all of us who live with animals and wish to know them as individual beings with mind and heart. In her words, I read reverence.

BEING ANIMALS

Perhaps an apt title for this book would have been not *Being with Animals*, but *Being Animals*. We humans are part of the natural world, not only in relationship with it. *We are part of the natural world.* How many times in the last few years has each of us heard this mantra, or read it? It cascades from the lips of those people passionate about the environment, about animal welfare, about let's-not-be-a-consumer-society-anymore. It's said by sky-gazers who look at the heavens and find kinship with the stars, by visitors to a symbol-using ape who find a soul mate.

But is it really so simple?

There's first of all the question of what's *natural*, and whether we would, or should, feel any less in relationship with a cloned puppy than with a "normal" one. Even aside from this, complexities abound. *We're part of the natural world.* Yes, we are 98 percent chimpanzee, in terms of genetic similarity. But as the anthropologist Jonathan Marks reminds us, measuring our chromosomal commonality with our closest living relatives does little justice to what we humans are all about. If humans share over 25 percent of our genes with a daffodil (as in fact we do), in what meaningful sense are we one-quarter flower?[15] Because we are biocultural in a sense that no other animal is, because we create whole visionary worlds for ourselves as no other animals do, I see what Marks means. He's right.

Yet when I delve into Donna Haraway's book *When Species Meet*, and encounter a chapter titled "We Have Never Been Human," I immediately grasp what she means (or at least I create a meaning for myself from Haraway's words!). Haraway celebrates her own non-humanness:

I love the fact that human genomes can be found in only about 10 percent of all the cells that occupy the mundane space I call my body; the other 90 percent of the cells are filled with the genomes of bacteria, fungi, protists, and such, some of which play in a symphony necessary to my being alive at all, and some of which are hitching a ride and doing the rest of me, of us, no harm. I am vastly outnumbered by my tiny companions; better put, I become an adult human being in company with these tiny messmates.[16]

Haraway doesn't become anything less by having DNA of bacteria, or DNA of apes, in her body; she becomes more. She becomes a mosaic of a person who transcends her evolutionary past even as she celebrates it.

Epilogue

•　　•　　•　　•

The flock grazing just beyond them was like
every such flock since the beginning of the world.
These might have been Abraham's sheep or David's,
returning to the fold at night, called forth in the morning,
each by name to pasture in the wilderness, coming at
midday to a stream of living water to drink.

—Sterling North, *So Dear to My Heart*

SHEEP GRAZE (and always have), sheep seek water (and always
have), sheep seek the safety of the flock (and always have). In this
way, animals embody a kind of timelessness. Even the world's most
attuned dog—a dog that predicts his favorite person's unexpected
arrival home, and then shares a walk with her—remains oblivious to
the tragedies and triumphs of the human world. Hints of this timeless
quality in animals appear in Verlyn Klinkenborg's book *The Rural Life*.
Recalling his train ride out of Manhattan, three days after the
September 11th terrorist attacks on New York and Washington,
Klinkenborg writes:

The world into which I was passing exuded nothing but its
own repose. It had no news to deliver, or rather only the old,

inarticulate news that bricks and water and steel have always
delivered.

After a while I reached my stop and drove north along a high-
way through the cornfields. Here too I felt the same thing, that
there was a mute voice in the extreme order of those rows of
corn. . . .

At home the horses and dogs consoled me in a way I couldn't
understand, until I finally realized that they could not be told
what had happened that week. In that fact lay the consolation.
They had only the old news to give, their old satisfaction with
the world as they know it.[1]

Raw with the emotions surrounding the 9/11 tragedy, Klinkenborg
left, for a while, the world of human concerns. In that other world
inhabited by horses, dogs, and other animals wild and tame, all of us
find solace and also a primal resonance. There, we discover the universe's own rhythm, a rhythm no more determined by human action
than is the Earth's flight around the sun.

What a relief and a release it can be to journey to this place. In this
high-tech century, we track our driving with GPS and our children
with cell phones and our pets with microchips. We manage our time
to the microsecond. Yet even as we connect to our computers, we
yearn to connect with other animals. That urge itself is timeless. We
may think robots are cute, but when we feel sick or downhearted, it's
not a robot we yearn for to lay its head on our lap. Only being with
animals takes us to a place out of time, a place where all living creatures dwell.

Notes

•   •   •   •

CHAPTER 1. A Holy Procession of Animals

1 Though speculative in its details, this reconstruction is strongly supported by information from archaeologists, including Steven Mithen in *The Singing Neanderthals* (London: Weidenfeld and Nicolson, 2005) and Brian Hayden in *Shamans, Sorcerers, and Saints* (Washington, D.C.: Smithsonian Institution Press, 2003).

2 N. Russell and B. S. During, "Worthy Is the Lamb: A Double Burial at Neolithic Catalhöyük (Turkey)," *Paléorient* 32, no. 1, pp. 73–84.

3 Mark Verhoeven, "Ritual and Ideology in the Pre-Pottery Neolithic B of the Levant and Southeast Anatolia," *Cambridge Archaeological Journal* 12 (2002): 233–58.

4 Karen Armstrong, *The Great Transformation: The Beginning of Our Religious Traditions* (New York: Alfred A. Knopf, 2006), pp. 295–96.

5 Peter Fimrite, "Daring Rescue of Whale off Farallones," *San Francisco Chronicle*, December 14, 2005.

6 Thomas Berry, "Prologue: Loneliness and Presence," in P. Waldau and K. Patton, eds., *A Communion of Subjects: Animals in Religion, Science, and Ethics* (New York: Columbia University Press, 2007), p. 8.

CHAPTER 2. Humans Emerging: From Savanna to Art Cave

1 Jean-Marie Chauvet, Eliette Brunel Deschamps, and Christian Hillaire, *Dawn of Art* (New York: Harry N. Abrams, 1996), p. 58.

2 Ibid., pp. 41–42.

3 Donna Hart and Robert W. Sussman, "Man the Hunted: Just Another Item

on the Menu," in *Physical Anthropology Annual Editions 08–09*, edited by Elvio Angeloni (Boston: McGraw-Hill).

4 Steven Mithen, "The Hunter-Gatherer Prehistory of Human-Animal Interactions," in *The Animals Reader*, edited by Linda Kalof and Amy Fitzgerald (Oxford, England: Berg, 2007), pp. 117–28.

5 Ibid., p. 119.

6 Ibid., p. 120.

7 Elizabeth Marshall Thomas, *The Old Way: A Story of the First People* (New York: Farrar, Straus and Giroux, 2006), p. 90.

8 Ibid, pp. 90–92.

9 Adam Leith Gollner, Scribner, 2008.

10 "Lower Palaeolithic Hunting Spears from Germany," *Nature* 385 (27 February 1997), 807–10.

11 http://www.cnn.com/2008/WORLD/asiapcf/05/29/india.elephant.ap/index.html?iref=newssearch

12 Hart and Sussman, "Man the hunted."

13 http://news.bbc.co.uk/2/hi/science/nature/2885663.stm

14 http://news.bbc.co.uk/2/hi/science/nature/3310233.stm

15 Marian Vanhaeren, Francesco d'Errico, et al., "Middle Paleolithic Shell Beads in Israel and Algeria," *Nature* 312 (2006): 785–88.

16 http://news.bbc.co.uk/2/hi/science/nature/3629559.stm

17 Here I exclude *Homo floresiensis*, the so-called Hobbit, which apparently lived longer than the Neanderthals in coexistence with modern humans—but that's another story.

18 I give more detail on Neanderthal behavior of this sort in chapter 4 of my book *Evolving God*.

19 Paul Shepard, *The Others: How Animals Made Us Human* (Washington, DC: Island Press, 1996), p. 11.

20 Take a virtual tour at http://www.culture.gouv.fr/culture/arcnat/chauvet/en/.

21 Jean Clottes, *Return to Chauvet Cave: Excavating the Birthplace of Art—The First Full Report* (London: Thames & Hudson, 2003).

22 Ibid., p. 149.

23 Ibid., p. 206.

24 Ibid., p. 202.

25 Phaidon Press, *30,000 Years of Art* (London: Phaidon Press, 2007).

26 Ibid., p. 15.

27 Michael Balter, "Paintings in Italian Cave May Be Oldest Yet," *Nature* 290 (2002): 419–21.

28 Paul G. Bahn, *Cave Art: A Guide to the Decorated Ice Age Caves of Europe* (London: Frances Lincoln Limited, 2007), p. 24.

29 This basic point is made well in J. D. Lewis-Williams and T. A. Dowson, "The Signs of All Times: Entopic Phenomena in Upper Palaeolithic Art," *Current Anthropology* 29, no. 2 (1986): 201.

30 Shepard, *The Others*, p. 91.

31 *30,000 Years of Art*, p. 14.

32 Bahn, *Cave Art*, p. 13.

33 http://findarticles.com/p/articles/mi_m1134/is_1_108/ai_53682803

34 Harold Morphy, *Aboriginal Art* (London: Phaidon, 1998).

35 I'm often asked about Julian Jaynes's *The Origin of Consciousness in the Breakdown of the Bicameral Mind* (Boston: Houghton-Mifflin Co., 1976) in this context. Though a compelling read, this book is decades old and quite out of date from a scientific point of view about what is known of our ancestors' cognitive powers.

36 Morphy, *Aboriginal Art*, pp. 68–69.

37 Peggy Reeves Sanday, *Aboriginal Paintings of the Wolfe Creek Crater* (Philadelphia: University of Pennsylvania Museum of Archaeology and Anthropology, 2007), p. 27.

38 Morphy, *Aboriginal Art*, p. 68.

39 Ibid., p. 23.

40 Ibid., p. 100.

41 Boria Sax, "Animals as Tradition," in *The Animals Reader*, Linda Kalof and Amy Fitzgerald, eds. (Oxford, England: Berg, 2007), p. 276.

CHAPTER 3. Taming the Wild?

1 Tom D. Dillehay, C. Ramirez, et al., "Seaweed, Food, Medicine, and the Peopling of South America," *Science* 320 (2008): 784–89, quoted material from p. 785.

2 M. Thomas, P. Gilbert, et al., "DNA from Pre-Clovis Human Coprolites in Oregon, North America," *Science* 320 (9 May 2008): 786–98.

3 Jennifer A. Leonard, R. K. Wayne, et al., "Ancient DNA Evidence for Old World Origin of New World dogs," *Science* 298 (2002): 1613–16.

4 Robert K. Wayne, Jennifer A. Leonard, et al., "Genetic Analysis and Dog Domestication," in *Documenting Domestication*, edited by Melinda A. Zeder, Daniel G. Bradley, et al. (Berkeley: University of California Press, 2006), pp. 279–93.

5 Peter Savolainen, Ya-ping Zhang, et al., "Genetic Evidence for an East Asian Origin of Domestic Dogs," *Science* 298 (2002): 1610–13.

6 Elizabeth Pennisi, "A Shaggy Dog History," *Science* 298 (2002): 1540–42.

7 This is my take on ideas found in Stephen Budiansky's *The Covenant of the Wild* (New Haven: Yale University Press, 1992).

8 Budiansky, *The Covenant of the Wild*, p. 20.

9 Michael Korda, "Thundering Hooves," *Washington Post BookWorld*, July 13, 2008, p. 9.

10 Paul Shepard, *The Others: How Animals Made Us Human* (Washington, DC: Island Press, 1996), p. 267.

11 Budiansky, *The Covenant of the Wild*, p. 61.

12 Darcy F. Morey, "Burying Key Evidence: The Social Bond Between Dogs and People," *Journal of Archaeological Science* 33: 158–75, p. 166.

13 Wayne, Leonard, et al., "Genetic Analysis and Dog Domestication."

14 Ibid., p. 289.

15 http://www.spiegel.de/international/world/0,1518,504508,00.html

16 http://www.eva.mpg.de/psycho/dogs/dogs_participation.html

17 Julia Riedel, Katrin Schumann, et al., "The Early Ontogeny of Human-Dog Communication," *Animal Behaviour* 75: 1003–14, quoted material from p. 1012.

18 Morey, "Burying Key Evidence," quoted material from p. 171.

19 Ibid., p. 166.

20 J-D. Vigne, J. Gullaine, et al., "Early Taming of the Cat in Cyprus," *Science* 304 (2004): 259.

21 http://www.cnn.com/video/#/video/offbeat/2008/07/15/kachroo.uk.pet.sheep.itn

22 N. Russell and B. S. During, "Worthy Is the Lamb: A Double Burial at

Neolithic Catalhöyük (Turkey)," *Paléorient* 32, no. 1: 73–84, quoted material from p. 76.

23 Nerissa Russell, "The Domestication of Anthropology," in *Where the Wild Things Are Now*, edited by Rebecca Cassidy and Molly Mullin (Santa Fe, NM: School of American Research Press, 2007), pp. 27–48, quoted material from pp. 39, 41–42.

24 Melinda A. Zeder and Brian Hesse, "The Initial Domestication of Goats (*Capra hircus*) in the Zagros Mountains 10,000 Years Ago," *Science* 287 (2000): 2254–57.

25 Some of my interpretation of Zeder and Hesse's work is derived from Curtis W. Marean, "Perspectives. Anthropology: Age, Sex, and Old Goats," *Science* 287 (2000): 2174–75.

26 Budiansky, *The Covenant of the Wild*, p. 60.

27 Peter J. Wilson, "Agriculture or Architecture? The Beginnings of Domestication," in Cassidy and Mullin, eds., *Where the Wild Things Are Now*, p. 107.

28 Ibid.

29 Budiansky, *The Covenant of the Wild*, e.g., p. 100.

30 Ibid., p. 102.

31 Ibid., p. 23.

32 J. Edward Chamberlin, *Horse: How the Horse Has Shaped Civilization* (New York: BlueBridge, 2006), p. 40.

33 Sandra Olsen, "Early Horse Domestication on the Eurasian Steppe," in *Documenting Domestication*, edited by Melinda A. Zeder, Daniel G. Bradley, et al. (Berkeley: University of California Press, 2006), pp. 245–69, quoted material from p. 261.

34 Ibid., p. 263.

35 See Craig Childs, *Animal Dialogues: Uncommon Encounters in the Wild* (New York: Little, Brown, 2007).

CHAPTER 4. Cat Mummies and Lion Symbols

1 http://www.democrats.org/a/2005/06/history_of_the.php

2 Mary Voigt, "Catalhöyük in Context," in I. Kuijt, *Life in Neolithic Farming Communities* (New York: Kluwer/Plenum, 2000), pp. 253–93.

3 Ian Hodder and Craig Cessford, "Daily Practice and Social Memory at Catalhöyük," *American Antiquity* 69 (2004): 17–40, quoted material from p. 20.

4 Ibid., pp. 21–22.

5 Ibid., p. 35.

6 Brian Hayden, *Shamans, Sorcerers, and Saints* (Washington, DC: Smithsonian Institution Press, 2003), p. 191.

7 Descriptions taken from Marc Verhoeven, "Ritual and Ideology in the Pre-Pottery Neolithic B of the Levant and Southeast Anatolia," *Cambridge Archaeological Journal* 12, no. 2 (2002): 233–58.

8 Nicholas Birch, "7,000 Years Older Than Stonehenge: The Site That Stunned Archaeologists," *The Guardian*, Wednesday, April 23, 2008.

9 Verhoeven, "Ritual and Ideology," p. 252.

10 Ibid.

11 Retrieved July 23, 2008, from website of the Israel Museum in Jerusalem, http://64.233.169.104/search?q=cache:ogNPH2H4v78J:www.imj.org.il/eng/exhibitions/2002/Sacred_Animals/index.html+sacred+animals+of+egypt&hl=en&ct=clnk&cd=5&gl=us.

12 Douglas C. McGill, "Met Says Its Popular Cat Is Probably Fake," *New York Times*, April 30, 1987.

13 Nora E. Scott, "The Cat of Bastet," *The Metropolitan Museum of Art Bulletin*, New Series 17, no. 1 (Summer 1958): 1–7.

14 Ibid.

15 McGill, "Met Says Its Popular Cat Is Probably Fake."

16 Retrieved July 23, 2008, see http://www.metmuseum.org/toah/hd/lapd/ho_56.16.1.htm.

17 http://www.albanyinstitute.org/info/exhibits/egypt.dogmummy.htm

18 Paul Shepard, *The Others: How Animals Made Us Human* (Washington, DC: Island Press, 1996), p. 180.

19 M. J. Magee, M. L. Wayman, and N. C. Lovell, "Chemical and Archaeological Evidence for the Destruction of a Sacred Animal Necropolis at Ancient Mendes, Egypt," *Journal of Archaeological Science* 23 (1996): 485–92.

20 Stonehenge information from "Probing Stonehenge," *Science* 320 (April 11, 2008): 159; Michael Balter, "Early Stonehenge Pilgrims Came from Afar, with Cattle in Tow," *Science* 320 (June 27, 2008): 1704–5; "Stonehenge Was a Place of Burial, Researchers Say," cnn.com, http://www.cnn.com/

2008/TECH/science/05/29/stonehenge.ap/index.html; James Owen, "Stonehenge Settlement Found," *National Geographic News*, January 30, 2007, http://news.nationalgeographic.com/news/2007/01/070130-stonehenge.html.

21 Cecile Callou, Anaick Samzun, and Alain Zivie, "A Lion Found in the Egyptian Tomb of Maia," *Nature* 427 (2004): 211–12.

22 For video and comments, see http://talkislam.wordpress .com/2006/03/25/the-lion-proclaims-allah/.

23 Laura Hopgood-Oster, *Holy Dogs & Asses: Animals in the Christian Tradition* (Urbana: University of Illinois Press, 2008).

24 Ibid., p. 49; both stories, according to Hopgood-Oster, come from the Gospel of pseudo-Matthew, an "extracanonical account" that may date to the eighth century CE.

25 For more on the *Imago dei*, see Wentzel van Huyssteen's *Alone in the World? Human Uniqueness in Science and Theology* (Grand Rapids, MI: William B. Eerdmans, 2006) and responses to it in *Zygon*, Spring 2008.

26 I continue to rely on Hopgood-Oster's account for these stories, and encourage readers to consult her book for her full analysis.

27 http://dsc.discovery.com/convergence/lionsden/lionsden.html

28 For details of the animal forms, and for quoted material, I consulted the Metropolitan Museum website, http://www.metmuseum.org/toah/hd/best/ho_22.58.1.htm.

29 Hopgood-Oster, *Holy Dogs & Asses*, p. 77.

30 Lucia Impelluso, *Nature and Its Symbols* (Los Angeles: The J. Paul Getty Museum, 2004).

31 Hopgood-Oster, *Holy Dogs & Asses*, p. 5.

CHAPTER 5. Animal Souls

1 Karen Armstrong, *The Great Transformation* (New York: Alfred A. Knopf, 2006), pp. 242–43.

2 Ibid., p. 244.

3 "Inherent Value Without Nostalgia: Animals and the Jaina Tradition," in P. Waldau and K. Patton, eds., *A Communion of Subjects: Animals in Religion, Science, and Ethics* (New York: Columbia University Press, 2007), pp. 241–429.

4 Verbatim from BBC website, http://www.bbc.co.uk/religion/religions/jainism/living/ahimsa_2.shtml.

5 Armstrong, *The Great Transformation*, pp. 279, 283.

6 Brian Edward Brown, "Environmental Ethics and Cosmology: A Buddhist Perspective," *Zygon* 39, no. 4 (2004): 885–900, quoted material from pp. 887, 886.

7 Based on articles on terra.wire and pbs.com.

8 For example, see http://www.shrinesf.org/; thanks to Sherman Wilcox for pointing me toward this painting.

9 Thanks to Ginger Zarske for acquainting me with the movie.

10 Brother Ugolino, *Little Flowers of St. Francis of Assisi*, translated by W. Heywood (New York: Cosimo Classics, 2007), quoted material from p. 42.

11 http://www.ivu.org/news/1–96/linzey.html.

12 Posted at www.washingtonpost.com in October 2007.

13 Allen and Linda Anderson, *God's Messengers: What Animals Teach Us about the Divine* (Novato, CA: New World Library, 2003), pp. 4, 10, 81–82.

14 All quotes taken from *Best Friends Magazine*, January–February 2008, pp. 34–35.

15 Personal communication via e-mail, January 11, 2008.

16 From www.pbs.org, documentary *Holy Cow*, see http://www.pbs.org/wnet/nature/holycow/hinduism.html.

17 Reuters online article, May 9, 2007, see http://www.pbs.org/wnet/nature/holycow/hinduism.html.

18 Quoted from *Best Friends Magazine*, January–February 2008, p. 33.

19 Gary Kowalski, *The Souls of Animals*, revised 2nd ed. (Novato, CA: New World Library, 1999), p. 140.

20 Maurice Friedman, *Martin Buber's Life and Work: The Early Years 1878–1923* (New York: E. P. Dutton, 1981), p. 15.

21 Ibid., p. 12.

22 Speech, January 19, 1990.

23 Kowalski, *The Souls of Animals*, p. 22.

24 Ibid., p. 143.

25 http://www.ncf.ca/freenet/rootdir/menus/sigs/religion/buddhism/introduction/precepts/precept-1.html

26 http://online.sfsu.edu/%7Erone/Buddhism/BuddhismAnimalsVegetarian/ Buddhism%20and%20Animal%20Rights.htm

27 Quoted in Ronald Epstein, "A Buddhist Perspective on Animal Rights," paper presented at animal-rights conference, San Francisco State University, March 29–April 1, 1990, available on the Internet.

28 Ian Harris, "'A Vast Unsupervised Recycling Plant': Animals and the Buddhist Cosmos," in Waldau and Patton eds., *A Communion of Subjects*, p. 209.

29 http://plato.stanford.edu/entries/hartshorne/

30 Genesis 1:26, Holy Bible, King James version.

CHAPTER 6. Ravens, Shamans, and Dogs Who Dream

1 Craig Childs, *The Animal Dialogues: Uncommon Encounters in the Wild* (New York: Little, Brown and Co., 1997, 2007); the raven story is on pp. 127–38.

2 Eric Mortensen, "Raven Augury from Tibet to Alaska," in P. Waldau and K. Patton, eds., *A Communion of Subjects: Animals in Religion, Science, and Ethics* (New York: Columbia University Press, 2007), pp. 423–36.

3 Ibid., pp. 425, 431.

4 I refer to Tooy Alexander by his first name because I am following Richard Price's practice in the book *Travels with Tooy: History, Memory, and the African American Imagination* (Chicago: University of Chicago Press, 2008).

5 E-mail from Richard Price, August 25, 2008, used with permission.

6 Price, *Travels with Tooy*, p. 45.

7 Ibid., p. 227.

8 Brian Morris, *Religion and Anthropology* (Cambridge, England: Cambridge University Press, 2006), p. 23.

9 This list is adapted from Mathias Guenther, "From Totemism to Shamanism: Hunter-Gatherer Contributions to World Mythology and Spirituality," in *The Cambridge Encylopedia of Hunters and Gatherers*, edited by Richard B. Lee and Richard Daly (Cambridge, England: Cambridge University Press, 1999), p. 426–33.

10 Edith Turner, "The Making of a Shaman: A Comparative Study of Inuit, African, and Nepalese Shaman Initiation," *The Global Spiral*, retrieved September 12, 2008, from http://www.metanexus.net/magazine/tabid/ 68/id/10606/Default.aspx.

11 Morris, *Religion and Anthropology*, p. 36.

12 Turner, "The Making of a Shaman."

13 E-mail communication from Richard Price, September 5, 2008.

14 Piers Vitebsky, *The Reindeer People: Living with Animals and Spirits in Siberia* (Boston: Houghton Mifflin, 2005), p. 32.

15 Ibid., p. 58.

16 Ibid., p. 11.

17 Ibid., p. 12.

18 Ibid., pp. 285–88.

19 Eduardo Kohn, "How Dogs Dream: Amazonian Natures and the Politics of Transspecies Engagement," *American Ethnologist* 34, no. 1 (2007): 3–24, quoted material p. 4. My information on the Runa comes entirely from Kohn's work.

20 Ibid., p. 13.

21 Ibid., p. 11.

22 Kofi Opuku, "Animals in African Mythology," in Waldau and Patton eds., *A Communion of Subjects*, pp. 351–59, quoted passages pp. 351–53.

23 Ibid., p. 356.

24 Quotation attributed to "Wilson Scott 2000," viewed March 15, 2008.

25 Marina Brown Weatherly, foreword to Joseph Epes Brown, *Animals of the Soul* (Scranton, PA: Element Books, 1993).

26 Vine DeLoria Jr., *God Is Red: A Native View of Religion*, 3rd edition (Golden, CO: Fulcrum Publishing, 2003), quoted material from p. 253.

27 See Shepherd Krech, *The Ecological Indian* (New York: W. W. Norton, 2000), for a full development of this argument.

28 See http://primatology.net/2008/10/13/anjana-the-chimpanzees-bond-with-two-white-tigers/.

29 For a lovely example, see the video clip at http://www.youtube.com/watch?v=bpblo-B6bdo.

30 Lawrence Buell, ed., *The American Transcendentalists* (New York: Random House/Modern Library, 2006), quoted material from p. 436.

31 Ibid., p. 450.

32 Ibid., p. 469.

33 This poem and all of Whitman's work is archived online at http://www.whitmanarchive.org/.

34 Michael Robertson, "Reading Whitman Religiously," *Chronicle of Higher Education*, April 11, 2008, pp. B6–B9.

35 Buell, *The American Transcendentalists*, p. 417.

CHAPTER 7. Of Whales and Tortoises

1 Peter Fimrite, "Daring Rescue of Whale off Farallones," *San Francisco Chronicle*, December 14, 2005.

2 Lynne Cox, *Grayson* (New York: Knopf, 2006).

3 Ibid., p. 14.

4 Ibid., p. 41.

5 Ibid., p. 84.

6 Ibid., pp. 140–42.

7 Margaret Drabble, *The Sea Lady* (Orlando, FL: Harcourt, Inc., 2007), pp. 340–41.

8 Both books are by Isabella Katkoff, Craig Hatkoff, and Paula Kahumbu, with photographs by Peter Greste (New York: Scholastic, 2006, 2007).

9 From *Nature* magazine, news@nature.com, archived at http://www.nature.com/news/specials/tsunami/index.html.

10 Hatkoff et al., *Owen and Mzee: The True Story.*

11 Ibid.

12 Hatkoff et al., *Owen and Mzee: The Language of Friendship.*

13 This post, dated April 1, 2006, is credited to Lisa Hunt of Boise, Idaho.

CHAPTER 8. Articulate Apes and Emotional Elephants

1 In *Journal of Consciousness Studies* 8 (no. 5–7), 2001, p. 299.

2 Kanzi now resides at the Great Ape Trust in Des Moines, Iowa. For more information on Kanzi, see later in this chapter and www.greatape trust.org.

3 See my chapter on patterned interactions in apes, in *Anthropology Beyond Culture*, edited by Richard Fox and Barbara J. King (Oxford, England: Berg Publishing, 2002).

4 For examples, see Jane Goodall, *Through a Window* (Boston: Houghton Mifflin, 1990).

5 For a full explication of this principle, see the articles in *The Journal of Developmental Processes*, at www.councilhd.ca.

6 F. Warneken et al., "Spontaneous Altruism by Chimpanzees and Young Children," *PloS Biology* 5, no. 7 (2007): 1414–20, quoted material is from p. 1417.

7 F. B. M. de Waal, "With a Little Help from a Friend," *PLoS Biology* 5, no. 7 (2007): 1406–8, quoted material is from p. 1406.

8 The Dalai Lama and Paul Ekman, *Emotional Awareness: Overcoming the Obstacles to Psychological Balance and Compassion* (New York: Henry Holt & Company, 2008).

9 Marc Bekoff and Jessica Pierce, *Wild Justice: The Moral Lives of Animals* (Chicago: University of Chicago Press, 2009).

10 S. Preston and F. de Waal, "Empathy: Its Ultimate and Proximate Bases," *Behavior and Brain Sciences* 25 (2002): 1–72.

11 Marc Bekoff, *The Emotional Lives of Animals: A Leading Scientist Explores Animal Joy, Empathy, and Sorrow—And Why They Matter* (Novato, CA: New World Library, 2007).

12 Readers who are interested in details may consult my books *The Dynamic Dance: Nonvocal Communication in African Apes* (Cambridge: Harvard University Press, 2004) and *Evolving God: A Provocative View on the Origins of Religion* (New York: Doubleday, 2007).

13 Christophe Boesch, *The Chimpanzees of Tai Forest* (Oxford, England: Oxford University Press, 2000).

14 William Mullen, "One by One, Gorillas Pay Their Last Respects," *Chicago Tribune*, December 8, 2004; zookeeper quoted is Amy Coons.

15 Bekoff, *The Emotional Lives of Animals*, pp. 128, 131.

16 Ibid., p. 64.

17 Jane Goodall, "The Dance of Awe," in *A Communion of Subjects: Animals in Religion, Science, and Ethics*, edited by P. Waldau and K. Patton (New York: Columbia University Press, 2007), pp. 651–56, quoted material from p. 654.

18 Ibid., p. 653.

19 See A. Whiten et al., "Cultures in Chimpanzees," *Nature* 399 (1999): 682–85.

20 Quoted material from http://www.gorilla-haven.org/ghsocialclimbing.htm.

21 I am grateful to my friend and colleague, the anthropologist Danielle Moretti-Langholtz, for alerting me to the *Nature* TV series segment about Shirley and Jenny, which started me down a path of reading that enhanced the elephant section of this chapter.

22 Unless otherwise noted, all quotes about elephants at the Sanctuary come from its website, accessed 7/5/07: http://www.elephants.com/.

23 G. A. Bradshaw, Allan N. Schore, Janine L. Brown, Joyce H. Poole, and Cynthia J. Moss, "Elephant Breakdown," *Nature* 433 (2005): 807.

24 Cynthia Moss, *Elephant Memories* (New York: William Morrow and Co., 2007), p. 270.

25 Ibid., pp. 270–71.

CHAPTER 9. Dog and Cat (and Buffalo) Mysteries

1 I first heard about Matt and Keith's dogs from Matt at a publishing dinner in November 2006, before I began to write this book. I followed up in August 2007 by talking with both Matt and Keith together.

2 Rupert Sheldrake, *Dogs That Know When Their Owners Are Coming Home* (New York: Crown/Three Rivers Press, 1999), p. 45.

3 Chris Dufresne, in Sylvia Browne and Chris Dufresne, *Spirit of Animals* (Cincinnati: Angel Bea Publishing, 2007).

4 Rupert Sheldrake, Terence McKenna, and Ralph Abraham, *The Evolutionary Mind: Conversations on Science, Imagination, and Spirit* (Rhinebeck, NY: Monkfish Publishing, 2005), p. 131.

5 Sheldrake, *Dogs That Know*, p. 25.

6 See A. Fogel, B. J. King, and S. G. Shanker, eds., *Human Development in the 21st Century* (Cambridge, England: Cambridge University Press, 2008).

7 "Grieving Couple Commits Suicide After Dog Dies," Reuters.com, April 2, 2007.

8 Interview by Deborah Solomon, *New York Times Magazine*, August 5, 2007.

9 http://www.time.com/time/nation/article/0,8599,1629962,00.html

10 Ibid.

11 http://72.14.253.104/search?q=cache:3sFszw1EwVYJ:www.redcross .org/news/ds/hurricanes/evacuationchecklist.html+Red+Cross+%2B +hurricane+katrina+%2B+pets&hl=en&ct=clnk&cd=2&gl=us

12 The full poem can be read, in many languages, here: http://
rainbowsbridge.com/Poem.htm. I have only seen the poem attributed to
"anonymous" and do not know an author to credit.

13 Stephanie V. Siek, "Grieving Pet Owners Find Solace in Sharing," *Boston
Globe*, March 30, 2006.

14 Unless otherwise noted, information on Charlie is taken from R. D.
Rosen's book *A Buffalo in the House* (New York: The New Press, 2007).

15 Ibid., p. 179.

16 Ibid., pp. 108–9.

17 Ibid., p. 107.

18 Ibid., p. 117.

19 See Sheldrake, *Dogs That Know*, p. 49, for a number of other possible explanations.

20 Sheldrake, *Dogs That Know*, p. 57.

21 See ibid., pp. 321–22, for debate about these results.

22 Sheldrake, *Dogs That Know*, p. 67.

23 Joanne Tanner, letter sent to *Cat Fancy* magazine, November 12, 1995.

24 The Sacred Depths of Nature (New York: Oxford University Press), p. 171.

CHAPTER 10. Finding Compassion

1 *Pig Tails*, PIGS Sanctuary newsletter, June 2006.

2 For an overview of Yellowstone's wildlife, geothermal features, and hiking
possibilities, try Tim Cahill's *Lost in My Own Backyard* (New York: Crown
Publishing Group, 2004).

3 Lee Whittlesey, *Death in Yellowstone: Accidents and Foolhardiness in the First
National Park* (Boulder, CO: Roberts Rinehart Publishers, 1995).

4 R. D. Rosen, *A Buffalo in the House* (New York: The New Press, 2007), p. 14.

5 See www.storyofthebison.com

6 Information in this paragraph is from Ken Zontek, *Buffalo Nation: American
Indian Efforts to Restore the Bison* (Lincoln: University of Nebraska Press, 2007).

7 Rosen, *A Buffalo in the House*, p. 19.

8 http://www.storyofthebison.com/tatankaaboutbison.asp

9 Zontek, *Buffalo Nation*, p. 144.

10 Black Elk, recorded and edited by Joseph Epes Brown, *The Sacred Pipe,* 2nd ed. (Norman: University of Oklahoma Press, 1989).

11 Ibid., p. 9.

12 Ibid., p. 9.

13 http://www.buffalomuseum.com/whitecloudealf.htm

14 Quoted material from Zontek, *Buffalo Nation,* p. 3.

15 Rosen, *A Buffalo in the House,* pp. 223, 231.

16 http://www.buffalofieldcampaign.org/faq/whyslaughter.html

17 Rosen, *A Buffalo in the House,* p. 227.

18 Zontek, *Buffalo Nation,* p. xiii.

19 http://itbcbison.com/index.php

20 http://itbcbison.com/recipes.php

21 Article by Marian Burros, *New York Times,* August 15, 2007.

22 "Update from the Field," e-mail message from the Buffalo Field Campaign, sent 12/27/07.

23 https://www.tankabar.com/cgi-bin/nanf/main/tankabar.cvw?sessionid =3f32a782441f4443f939d2d01c9cc2ece055cb9

24 http://www.sciencedaily.com/releases/2006/01/060106002944.htm

25 "Therapeutic Use of Companion Animals in Health Care," *Journal of Nursing Scholarship* 29, no. 3: 249–54, quoted material is from pp. 253–54.

26 Gail F. Melson, *Why the Wild Things Are: Animals in the Lives of Children* (Cambridge: Harvard University Press, 2001). I cannot say whether the diagnoses were accurate or whether the boys' confinement at a residential treatment facility was wise. As an anthropologist, I wonder about the medicalization of aggression in children; but that is another story.

27 Ibid., p. 120.

28 Meg Olmert, *Made for Each Other: The Biology of the Human-Animal Bond* (Cambridge: Da Capo, 2009).

29 I write a monthly column, often about animal books, for the online magazine Bookslut. My review of *Made for Each Other* is at http://www .bookslut.com/features/2009_01_014021.php

30 Gail F. Melson, "Children in the Living World: Why Animals Matter for Children's Development," in *Human Development in the 21st Century,* edited by A. Fogel, B. J. King, and S. G. Shanker (Cambridge, England: Cambridge University Press, 2008), p. 149.

31 Ibid., p. 151.

32 Ibid., p. 152.

33 Ibid., p. 153.

34 http://www.worldwithoutus.com/did_you_know.html

CHAPTER 11. Clones, Crows, and Our Future

1 W. Miller et al., *Nature* 456 (2008): 387–90.

2 Henry Nicholls, "Let's Make a Mammoth," *Nature* 456 (2008): 310–14.

3 For more on babies and learning to breathe, see the work of James McKenna on co-sleeping, as represented here: http://www.nd.edu/ ~jmckenn1/lab/faq.html.

4 "Bring Back the Woolly Mammoth?" *New York Times*, November 23, 2008.

5 To understand more about the science of Dolly, see my review of Sarah Franklin's book *Dolly Mixtures* at http://www.bookslut.com/features/ 2008_03_012498.php.

6 William Saletan, "The Fruit of Our Sirloins," *Washington Post*, January 7, 2007.

7 Yasmine Phillips, "Cloned Food 'On Shelves' in 2–3 Years," *The West Australian*, October 25, 2008.

8 Jim Mason and Mary Finelli, "Brave New Farm?" in *In Defense of Animals: The Second Wave*, edited by Peter Singer (Oxford, England: Carlton Blackwell, 2006), pp. 104–22.

9 See "Company Cloning Pet Dogs, for Hefty Fee," February 19, 2008, at CBS News: http://www.cbsnews.com/stories/2008/02/19/earlyshow/ living/petplanet/main3843862.shtml.

10 "Beloved Pets Everlasting?" *New York Times,* December 31, 2008, at http://www.nytimes.com/2009/01/01/garden/01clones.html ?pagewanted=1&_r=1&emc=eta1.

11 http://www.nopetcloning.org/images/pressrelease_petcloning _report.pdf

12 Andrew Linzey, *Animal Theology* (Champaign, IL: University of Illinois Press, 1994), p. 138.

13 Paul and Anne Ehrlich, *The Dominant Animal* (Washington, DC: Island Press, 2008).

14 My thanks to Nuala Galbari, who has given me permission to use information sent in e-mail communications with me on December 9, 2008.

15 See http://personal.uncc.edu/jmarks/interests/aaa/marksaaa99.htm, or Jonathan Marks, *What It Means to Be 98% Chimpanzee* (Berkeley: University of California Press, 2003).

16 Donna J. Haraway, *When Species Meet* (Minneapolis: University of Minnesota Press, 2008), pp. 3–4.

EPILOGUE

1 Verlyn Klinkenborg, *The Rural Life* (Boston: Little, Brown and Company, 2002), pp. 159–60.

Acknowledgments

● ● ● ●

To everyone who talked with (or wrote to) me about animals, a big thank you: Anna Autilio; Karen Flowe and Ron Flowe; Nuala Galbari and David Justis; Charles Hogg; Matt Hurley and Keith Lustig; Reverend Steve Keplinger; Richard Price; Joanne Tanner; and Ginger Zarske.

Lisa Stevens at the Smithsonian's National Zoological Park and Sue Savage-Rumbaugh at the (then) Language Research Center at Georgia State University gave me wonderful opportunities to observe African apes up close, and shared some of their knowledge with me. I am very grateful.

At the College of William and Mary, many people have supported and inspired me. My faculty colleagues Joanne Bowen, Matthew Liebmann (now at Harvard University), Danielle Moretti-Langholtz, Mary Voigt, and Brad Weiss made me think better about animals. Dean Carl Strikwerda uttered four perfect words to me at just the right time in my academic career. Joseph McClain writes like a dream and knows how to make a fellow writer look good. My primatology students, through their continuing dedication to science and to animal welfare, will change animals' futures for the better.

At Doubleday, Trace Murphy realized what I needed to do all along (Trace, you were right) and helped me see how to do it. Darya Porat saved my sanity more than once with her organization and persistence. With characteristic humor and kindness, James Levine and Lindsey Edgecombe at the Levine-Greenberg Literary Agency offered acutely insightful advice on the manuscript, as well as practical help from proposal through publication. You're the best.

Special hugs to Elizabeth King, my mom, who gave me books and cats right from the start, and who shares so much with me now; to Stuart Shanker, for unwavering support, for being brilliant (you'll hate me saying that but it's true), and for so many conversations over so many years; to Stephen Wood, for help in many bookish ways, for the sacrifice of sea foam and other adventures in honor of the greater good, and for these thirty-five years; and to all my dear friends near and far who sustain me with humor, books, and lots of chocolate.

Words cannot easily convey how much inspiration I take from the people who care for animals in need, day by day—when they are tired, when the rain is torrential and the humidity is terrible, when it's just too discouraging. My personal heroes include Marc Bekoff, Jane Dewar, Michael Fay, Nuala Galbari, Jane Goodall, David Justis, Tonya Higgins, Charles Hogg, Lyn Layer, Ginger Zarske, everyone at Best Friends Sanctuary in Utah, caretakers at animal sanctuaries like Save the Chimps and The Center for Great Apes (both in Florida), Pigs Sanctuary (in West Virginia), and indeed the caretakers at all the well-run animal sanctuaries throughout the world.

To Charles Hogg and Sarah Hogg, and to all the animals who share our lives, I offer the heartfelt hope that each of you in your own way feels the truth: you make my every day a happiness.

Index

• • • •